THEY POISONED THE WORLD

Life and Death in the Age of Forever Chemicals

Mariah Blake

CROWN
NEW YORK

Crown
An imprint of the Crown Publishing Group
A division of Penguin Random House LLC
crownpublishing.com

Library of Congress Cataloging-in-Publication Data
is available upon request.

Hardcover ISBN 978-1-5247-6009-0
International edition ISBN 978-8-217-08709-9
Ebook ISBN 978-1-5247-6011-3

Editor: Amanda Cook
Editorial assistant: Katie Berry
Production editor: Joyce Wong
Designer: Aubrey Khan
Production: Heather Williamson
Copy editors: Patricia Boyd, Sigi Nacson
Proofreaders: Christopher Pitts, Ethan Campbell
Indexers: J S Editorial, LLC, Eldes Tran
Publicist: Penny Simon
Marketer: Chantelle Walker

Manufactured in the United States of America

First Edition

THEY POISONED THE WORLD

In memory of Doc,

whose legacy will reach down

through generations

Contents

Author's Note

This is a work of narrative nonfiction. Most of the reporting took place between 2015 and 2024. Unless otherwise noted, the events and dialogue from that period were witnessed firsthand by the author or reconstructed from interviews, emails, text messages, social media posts, and personal diaries. The descriptions of historical events were pieced together from thousands of disparate sources, including archival records, court papers, contemporaneous news accounts, and U.S. government documents, many of them previously classified.

Preface

On the night I learned I was pregnant with my son, in late 2009, I dreamt I had been swept out to sea by a tsunami. I woke up with my heart racing, my elation at the day's news subsumed by a welling anxiety. This feeling would trail me through my early pregnancy, as I wrestled with how best to keep my child safe. I dealt with it in the same way I deal with many things: by burying myself in research.

What I learned was both awe-inspiring and sobering. In the months before a baby's birth, the umbilical cord pumps gallons of blood back and forth to the womb, delivering the nutrients and oxygen that the fetus needs to thrive. In the past, scientists believed the placenta filtered out most damaging pollutants passed on from the mother, but that is not the case. During those crucial early weeks when the cells knit themselves into brain and organs and fingers and lips, a steady flow of man-made chemicals pulses through the umbilical cord, permeating the fetus's blood and tissue. This is why, from their very first moments of life, every American newborn carries a slew of synthetic chemicals in their body.[1]

To my astonishment, I also learned that many of these substances have *never* been tested for safety. Of those that have, a substantial share are known to be damaging. Some can cause cancer or retard fetal development. Others alter the levels of hormones in the womb, causing subtle changes to a baby's brain and organs. Though these changes may not be apparent at birth, they can lead to a staggering variety of problems, including cancer, heart disease, infertility, early puberty, reduced IQ, and neurological disorders like ADHD. The evidence is so overwhelming that major medical organizations now warn that widespread exposure to hormone-disrupting substances may be a key factor behind alarming health trends, from plummeting sperm counts to rising rates of obesity and diabetes.[2]

Needless to say, for the rest of my pregnancy, I did my best to steer clear of toxic chemicals. I rummaged through my cupboards, weeding out plastic dishes and food containers, which often contain the infamous hormone-altering chemical BPA. After learning there were traces of toxic chemicals in my local water supply, I also ordered bottled water delivery. One snowy afternoon, a burly man in a Deer Park uniform schlepped a plastic cooler up the stairs to my apartment, along with three large jugs of spring water. Every day from that point on, I drank water from the cooler and used it to brew my morning coffee. After my son was born, I mixed his baby cereal with it and poured it into his stainless-steel sippy cup, all the while presuming it was safer than the water flowing from my tap.

I can't recall exactly how or why the doubt crept in. But around the time my son turned three, I began to worry that the cooler bottles might contain BPA. There was nothing about it on the labels, so I called Deer Park to find out. The answer left me gutted. In my efforts to protect my son, I had exposed him to a substance that might already have sown seeds of disease in his body.[3]

I found myself wondering why a dangerous chemical like BPA was even allowed in water bottles. Doing some more reading, I came across a deeply troubling study that had recently appeared in a major scientific journal. The authors had analyzed hundreds of "BPA-free" plastic food containers and packages and found that "almost all" of them leach chemicals with the very same properties that made BPA so harmful: They alter the levels of the hormone estrogen in our bodies. Strikingly, a leading U.S. plastic manufacturer whose BPA-free products had become a staple for health-conscious brands like Nalgene and Whole Foods had sued to block the researchers from publicizing their findings.[4] My interest was piqued. Delving into the court record, I unearthed something even more sinister.

It turned out the plastics industry was well aware that many products marketed as safe BPA-free alternatives actually release other damaging chemicals. For years, corporate scientists had been studying the problem and burying their own damning findings. At the same time, the industry had worked to cast doubt on research from outside scientists—often employing the same methods and consultants that the tobacco industry had used so successfully to discredit the science on smoking.

I eventually distilled some of my findings into an article that appeared in *Mother Jones* magazine in 2014.[5] After it was published, I was flooded with emails from worried parents who wanted to know what they could do to protect their families. I had no idea what to tell them. Under our current regulatory system, manufacturers don't have to disclose which chemicals they're using, much less test those chemicals for safety. Meaning that in a world overflowing with synthetic materials, people have no way of knowing which ones might be harmful.

As a journalist and a mother, I wanted to understand how Americans had wound up as lab rats in this foolhardy chemistry experiment. So I kept digging. I interviewed historians, tracked down government whistleblowers, and burrowed deep into archives, tracing the roots of the story to luminaries like Thomas Jefferson and Albert Einstein and piecing together the details of a secret

Manhattan Project chemistry program that would prove almost as revolutionary as the creation of the atomic bomb, though it barely registered in the history books.

These early forays led to several eye-opening realizations. First, the American chemical industry and the man-made materials that pervade our lives owe their existence largely to the U.S. government, particularly the centuries-long partnership between the government and the chemical giant DuPont. Second, chemical interests had not merely borrowed tactics from cigarette makers. On the contrary, they had *invented* many of the methods that Big Tobacco and other industries would later deploy to pick apart the science tying lucrative products to disease. It was a DuPont-funded scientist who first articulated the principle that now forms the bedrock of our system for regulating potentially toxic substances—namely, that they should be presumed safe until proven otherwise.

I felt this history, much of which has never been told, was key to understanding how we had arrived at this place as a nation. A place where mistrust of science hobbles our response to mortal threats like climate change. A place where simple acts like wearing masks to avoid deadly viruses are hopelessly politicized, where large swaths of the public spurn objective fact and impartial institutions, leading to an ever-more-divided society. A place where every child is born pre-polluted.

I decided to write a book that told this story through a single class of chemicals. Known as per- and polyfluoroalkyl substances, or PFAS, they offer a particularly potent illustration of both the promise and peril of synthetic materials. Their remarkable properties helped usher in the era of space travel and high-speed computing and paved the way for lifesaving medical breakthroughs. They also transformed thousands of everyday items—dental floss, clothing, furniture, food packaging, and carpet, to name just a few.

On the other hand, PFAS are extremely toxic, even in the most minuscule doses. They refuse to break down in nature, which is why they're commonly called forever chemicals. And they spread rapidly through air and water, polluting plants and wildlife around the world. They suffuse the blood of polar bears in the Arctic, eagles in the American wilderness, and fish in the deepest reaches of the ocean. They permeate snow on Mount Everest and fall with rain around the world.[6]

They also pollute the bodies of virtually every person on the planet.[7] Once inside us, they stay there like a ticking time bomb of disease. The best studied of these chemicals have been linked to obesity, infertility, testicular cancer, thyroid disease, neurological problems, immune suppression, and life-threatening pregnancy complications, among numerous other maladies.[8]

And yet the world probably wouldn't even know that they existed if it weren't for a family of West Virginia farmers. In the 1990s, after cattle they were grazing near a local DuPont landfill started dying off, the family sued the company. Their case wound up exposing a breathtaking cover-up involving the forever chemical PFOA, which DuPont used to make Teflon. It also inspired a frenzy of scientific research and a major class-action lawsuit.[9] But the magnitude of what these rural Appalachian farmers had uncovered hardly registered with the national media or the general public. When I began my research, almost no one outside scientific circles had heard of PFOA, much less the thousands of other forever chemicals that are so ubiquitous in American life.

That began to change in early 2016, when a contamination crisis in the upstate New York village of Hoosick Falls burst into the national headlines.[10] As in West Virginia, the problem was brought to light by ordinary citizens. In this case, it began with a young man named Michael Hickey, who had started questioning the safety of the local drinking water after losing several friends and relatives to cancer.

When I traveled to Hoosick Falls to meet Michael, I was struck by how ill-suited he seemed to activism. A clean-cut insurance underwriter with a fear of public speaking, he had no interest in environmental issues generally. He liked to joke that he got his news from ESPN. And yet somehow, he was spearheading a fight against several giant multinational corporations and government agencies to get his community clean drinking water.

Through Michael, I met other area residents who were enmeshed in the same battle. A beloved doctor who had documented unusually high rates of rare, aggressive cancers among his patients. A young mother who had put everything she had into a dream home for her family, only to learn that her private well was poisoned. A high school music teacher who decided to run for public office after getting some awful news about his youngest child. Like Michael, these were people who had spent their lives avoiding politics, trusting that there were systems in place to protect them. Now, they were utterly consumed by the struggle to protect their families and community from a chemical that the people at the highest levels of industry and government had long known was harmful. I decided to spend more time in Hoosick Falls and see where their struggle led.

Over the next eight years, I accompanied them to courthouses, political gatherings, hospital rooms, churches, dive bars, Little League games, and family dinners. I watched as they fought illness and grief. I also witnessed deeply emotional moments, including the birth of a child whose mother was all too aware of the chemical burden she had passed on through her umbilical cord.

As the crisis spread, first to neighboring towns and then to hundreds of municipalities across the country, I saw them band together with people from other affected communities, eventually taking their fight all the way to Capitol Hill. In the process, they touched off a spectacular chain reaction that would lead to dozens of congressional hearings, an avalanche of media coverage, and hundreds of bipartisan bills cropping up in Washington and statehouses nationwide. They also helped fuel a sprawling legal battle that would dwarf even legendary litigation like the case between U.S. states and Big Tobacco. This book tells their story.

Their journey has been extraordinary, but it also illuminates a problem facing every American. We live in a synthetic world. Our homes and workplaces are brimming with man-made materials. Our bodies are saturated with their chemical residue. The implications for public health are hard to overstate. But we can't count on our leaders to protect us from these threats without intense, sustained public pressure. The companies advocating inaction are too powerful, the systems favoring it too entrenched. It is up to us to protect ourselves and make a safer future for our grandchildren and great-grandchildren, who will reap the consequences of the choices we make now.

1

A Bump in the Road

The morning of October 5, 2010, started as most mornings did in the Hickey household. Ersel Hickey, a sixty-eight-year-old school bus driver in the village of Hoosick Falls, New York, rose before dawn. He pulled on his jeans and baseball cap and dropped by Stewart's convenience store for some coffee. Then he headed out for his morning shift. Ersel's bus route wound through the village's Second Ward, a mix of old brick factories and faded Victorian homes with American flags flapping from their porches, and then into hardscrabble hill country sprinkled with rusty trailers and ramshackle farmhouses. Along the way, Ersel cracked jokes, handed out candy, and let the kids blare music over the bus's radio.

On this particular day, after delivering the kids to school, Ersel returned to the brown-shingled cottage he shared with his wife, Sue. He grabbed a snack, slipped into the bathroom to relieve himself—and watched the toilet fill with blood. Ersel hollered for Sue, who gasped at the sight of the reddish water, then he headed straight to the office of the family's longtime doctor, Marcus Martinez.

Marcus examined him, took some X-rays, and decided to put

him in the hospital overnight for some tests. The doctor was on the phone making arrangements when Ersel's youngest son, Michael, dropped by the office to check on him. Marcus cupped his hand over the receiver and told Michael not to worry—Ersel was probably just constipated. After the inevitable quips about what Ersel was full of, Michael drove him to the hospital.

Michael, a thirty-one-year-old insurance underwriter with a firm belief in the power of positive thinking, spent the following day in his cubicle. Afterward, he attended a breastfeeding class with his fiancée, Angela. She was nine months pregnant with their first child, and Michael had been trying to persuade her to name the baby Ersel, so far without success. When he called his mother after class, he could tell from her voice that something was wrong. "I didn't want to tell you this over the phone," Sue told him. "But it's worse than what we thought. It looks like your father has kidney cancer."

Michael was stunned; it was hard to imagine someone so vigorous being seriously ill. But he was determined not to dwell on worst-case scenarios. The following night, he and Angela took Ersel out to dinner and Michael assured his father that everything would be okay. Ersel would get through the treatment, and then he could live out his dream of traveling the country in his RV. "This is just a bump in the road to the places you want to go," Michael told him.

At one point, Angela left the table, and Michael couldn't resist the urge to make a joke. "Maybe we can even sell her on Ersel now," he said, leaning in close to his father. "You know—maybe we can get something positive out of this situation." Michael had only seen his father cry once, but at the mention of his future grandson, Ersel started quietly sobbing.

About two weeks later, tiny blood clots began breaking off the tumor in Ersel's kidney, clogging his urinary tract. He was transported to St. Peter's Hospital in nearby Albany to have his kidney removed. Angela, who by then was past her due date, arranged to have her labor induced at St. Peter's, so the entire family could be there for Ersel's surgery.

The day of Ersel's procedure, she and Michael rose around 4 a.m., packed some baby clothes, and drove toward Albany through the cool autumn dawn. When they arrived at St. Peter's, Sue was sitting in the deserted waiting room with Michael's sister, Katy, their brother, Jeff, and Ersel's brother, Rich. Michael tried to distract himself with *Sports Illustrated,* but after an hour or two, he found himself fixating on an LCD screen with a scrolling color-coded list of patients. Navy blue meant pre-op; fuchsia meant undergoing surgery; lavender meant the patient was in recovery.

Gradually, the waiting room filled up. People in scrubs hustled back and forth through the swinging doors leading to the surgery suites. Patient after patient flipped from navy, to fuchsia, to lavender and then drifted off the screen. But Ersel's entry lingered up there in fuchsia. Finally, around two that afternoon, the surgeon emerged. The procedure had been more complicated than expected, he said. But Ersel had come through just fine—and he was now cancer-free. Exhausted but relieved, the family paid a brief visit to Ersel, who was lying unconscious in a recovery bay. Then Michael and Angela went down to the labor and delivery floor so that she could give birth to their son.

The Hickeys spent much of the following morning shuttling between Angela's and Ersel's rooms while Angela's relatives played rummy in the Pepto-Bismol-pink birthing suite. By early afternoon, Angela's contractions were coming hard and fast. But she wasn't making any progress, and after six or seven more hours of intense labor, the baby's heart rate began wavering. Her doctor was preparing for a cesarean when suddenly the baby entered the birth canal and emptied the contents of his bowels, a substance that can be fatal to a newborn if inhaled. The doctor ordered everyone except Angela's parents and Michael out of the room.

What happened next was a blur; nurses scurrying and the tap-tap of Angela's father's shoes as he paced behind the curtain they'd pulled around her bed. Then, around 10 p.m., Angela gave one last push and the baby emerged. Her mother cut the umbilical cord; a breathing specialist declared him healthy. The relatives who had been waiting in the hallway poured back into the room, many of them in tears. Michael was shaking so hard he had to sit down. Then Katy laid the swaddled newborn in his arms, and Michael held him close, marveling at his plump cheeks and tiny cleft chin.

Later, when the baby was sleeping, Michael asked the nurse to make an extra copy of the birth certificate. He took it to his father's room and pinned it on a board by the bed so Ersel could see the baby's name: Oliver Ersel Hickey. "You get to go home healthy, and Angela agreed to put Ersel in the middle, so it's a win all around," Michael told him. Ersel was so hazy from the surgery that he could barely open his eyes, but a smile flickered across his face.

A LOT OF THINGS CHANGED after Ersel's surgery. He'd always been a hard worker—for most of Michael's life, he'd not only driven the school bus but also worked the night shift at a local factory. He rarely complained, but his punishing schedule had kept him from doing things he dreamed of, like hitting the road in his RV or taking the family to the South Dakota Badlands.

Now, it seemed, he finally had the time. After the surgery, Ersel found driving the bumpy back-country roads on his bus route painful, so he moved to a part-time schedule and lavished attention on his new grandson. At one point, Oliver's sitter had to take several months off, and Ersel, who had rarely changed a diaper before Oliver's birth, volunteered to fill in. Each morning, he would arrive at Michael and Angela's with his newspaper and his Stewart's coffee, then spend the day playing on the floor with his grandbaby or proudly ferrying him around town in his stroller.

When Ersel wasn't with Oliver, he played golf or drove his buddies to the stock car races in his yellow Corvette. As usual, come spring, he rolled the loose change he'd collected all year and put it in a pail for gas money. Then he and Sue drove the RV to Myrtle Beach, South Carolina, where many blue-collar Hoosick Falls families vacationed. Sue, who had nerve damage in her legs and walked with a cane, dreaded these excursions. "I've always felt I deserve, once a year, on vacation, to have somebody else make my bed instead of me," she grumbled. But Ersel loved bumping into neighbors on the golf course and doing crosswords while they splashed in the pool with their families.

That June, following a routine scan, the hospital called Ersel back in for an extra test. Nobody thought much of it—they were too busy preparing for the big high school graduation party the family was throwing for Sue and Ersel's eldest grandchild. The day of the event, Ersel arrived at Katy's house with poster-sized photos of his grandson playing basketball. As the crowd gradually swelled, Ersel made the rounds like the local mayor—cracking jokes, talking sports, and asking after everyone's relatives. At some point, Marcus made a brief appearance, which struck Sue as strange. "He was in and out in a flash, and something about it gave me an awful feeling," she recalled. But Ersel didn't seem to notice. Around 11 p.m., when the party had dwindled to family and close friends, he lit some Chinese lanterns he'd brought back from South Carolina and stood mesmerized as the glowing orbs drifted into the night sky.

The following morning, Ersel got his scan results: The cancer was back, this time in one of his lungs. Heartsick, the Hickeys gathered on Sue and Ersel's porch and resolved to enjoy whatever time they had left together, as a family. Right away, Michael started planning a trip to the South Dakota Badlands. And two weeks later, all fourteen members of the Hickey clan boarded a plane wearing orange "Hickey Family Vacation" T-shirts—orange being the color of kidney cancer awareness, as well as Ersel's favorite football team, the Cleveland Browns. Ersel was enchanted with the otherworldly Badlands landscape and the colorful landmarks of the Old West. After visiting the Deadwood saloon, where "Wild Bill" Hickok was shot dead during a poker game, Ersel insisted on pushing Oliver's stroller up the hill to the cemetery to visit the legendary gunslinger's grave.

Not long after their return, Michael joined Ersel on his annual road trip with his golf buddies, as he had done for the past few years. The two men could hardly have been more different. Ersel was a boisterous type who could strike up a conversation with anyone. Michael was a chronic mumbler, often self-conscious at social gatherings.[1] He liked designer labels and fancy European cars. Some people around town dismissed him as money hungry, but Ersel's buddies treated Michael like family. While Ersel drove the RV, they all sat in the back, drinking beer and playing pitch. The men spent the rest of the weekend golfing near Syracuse. Despite his enthusiasm for the sport, Ersel was a notoriously bad player, and it wasn't unusual for him to hurl a club or a few choice profanities if a game wasn't going his way. During this trip, though, nothing seemed to faze him. Michael had the feeling that his father was just happy to be there at all.

That August, Ersel, Katy, and Sue made the three-hour drive to Beth Israel Deaconess Medical Center in Boston for Ersel's first week-long treatment with a powerful new drug. After that, Ersel and Sue spent every other week in Boston. Michael and Katy rearranged their work schedules so that one of them was always there, too. At first, the side effects were mild. Although Ersel was often tired, during off weeks he still played golf, doted on Oliver, and made his daily pilgrimage to Stewart's for coffee and lottery tickets.

But as the treatment progressed, Ersel developed blood-pressure problems, intense joint pain, and a case of thrush so bad he couldn't swallow. Michael, who had never considered himself much of a student, started obsessively researching kidney cancer in the hopes of finding a treatment that would ease his father's suffering. Between that and the Boston trips, he didn't have much energy left for his fiancée and son, but Angela was sympathetic. "All of us were willing to do what it took to get Ersel through this," she said. "We all thought this was going to be a stepping-stone and once Ersel was better, everything would go back to normal."

THE HIGHLIGHT OF ERSEL'S YEAR had always been his annual Christmas shopping trip. An all-day affair, it combined two of the things he loved most: group excursions and choosing gifts. Usually, he, Katy, Michael, and some other relatives would drive down to Crossgates Mall in Albany, and Ersel would hand everyone an envelope stuffed with cash to buy presents for their families. Then he'd make a beeline for Hannoush Jewelers, an elegant family-owned enterprise next to Forever 21, to find something special for Sue. After that, Ersel and Katy would do a round of speed shopping: sports gear for the grandkids, scarves for the women who worked at the local credit union where he was a board member. Some years, he wrote up a list of every child on his bus route and bought them presents, too. When the shopping was done, Ersel would take everyone out for a nice dinner and some spirited conversation—his favorite part of his favorite day.

Michael and Katy were determined to continue the tradition for their father's sake. So, in early December, they drove down to Crossgates with Angela and Ersel, who had just finished his final treatment. This time, he'd been too weak to handle more than a few rounds of the infusion. At the mall, he picked out a diamond necklace for Sue, but that was all he could manage. On Christmas Day, Ersel—who usually woke up early to cook breakfast and hand out the presents—slumped in his recliner, defeated.

By the following month, the pain in Ersel's knees and hips was so excruciating that he couldn't make it the dozen or so feet from the kitchen to the living room without using a walker. His appetite vanished, and he began sleeping most of the day. During these weeks, it seemed like half of Hoosick Falls cycled through the house to drop off food or visit with Ersel. Golf buddies, people he'd served with on the village board, parents who'd once ridden his bus and whose kids had ridden it, too. Sometime in January 2013, one of the organizers of the St. Patrick's Day parade came to inform Ersel he'd been picked as that year's grand marshal, one of the most coveted honors in the Irish Catholic village. Ersel was ecstatic. "All we kept hoping for, because he wanted it so bad, was for him to make it to the parade," Katy said.

By then, though, Marcus had begun to suspect that the cancer had spread. In early February, he ordered a PET scan. Katy happened to be standing in her parents' kitchen the following morning when Marcus walked in to deliver the results: The cancer had moved to Ersel's bones. When Katy asked whether her father would make it another six weeks to St. Patrick's Day, Marcus grimaced and shook his head. Then he stepped into the bedroom to tell Ersel.

2

Teflon Town

For much of its history, Hoosick Falls was best known as the home of Grandma Moses, a retired farmhand turned unlikely darling of the art world. Born in 1860, Anna Mary Robertson Moses took up painting in her seventies after her husband had died and her hands were too crippled from arthritis for embroidery. Sitting on her old swivel chair in her kitchen, she conjured "old-timey" pastoral scenes—vivid landscapes filled with joyful depictions of people working the land or taking part in seasonal rituals like *Apple Butter Making* and *Catching the Thanksgiving Turkey*.[1]

Moses's early work didn't draw much attention. When she entered some of her pictures in the county fair alongside her canned fruit and raspberry jam, only the preserves brought home ribbons. But in 1938, a Manhattan art collector passing through Hoosick Falls happened to see some of her paintings gathering dust in a drugstore window. Before long, her work was being displayed in the Museum of Modern Art and reproduced on postage stamps and Hallmark cards by the tens of millions. Norman Rockwell put her in one of *his* paintings. President Harry Truman invited her to Washington, calling her one of the "most popular and original painters" in America.[2]

A 1953 *Time* magazine cover featured an image of Grandma Moses standing on a snowy country road in a heavy, lace-collared dress, her silver hair swept up in a librarian bun, her sprightly gray eyes peering in wonder at something just beyond the frame. The accompanying story opened with a description of the landscape outside her home just north of Hoosick Falls—"the fallow corn and tomato fields falling away to the Hoosic River," the hills beyond "quilted with thick-ranked birch and maple trees and patches of frosty pasture land." It was this landscape that inspired and sustained her. Generations of Hoosick Falls residents would see their heritage in her paintings and take pride in her enormous popularity.[3]

But there was another aspect of the area's history that wasn't visible in her nostalgic images: its deep industrial roots. When the village was incorporated in 1827, it had just two hundred residents; the only mail service was a young boy who traveled by foot from a neighboring village. Then, in the 1850s, an entrepreneurial blacksmith broke ground on a massive farm-equipment factory with its own power plant and train station. The population swelled to more than seven thousand. Stately Victorian homes sprang up around the village center, along with department stores, an opera house, and a trolley system. The factory shut down just after World War I.[4] But the village's industrial infrastructure eventually attracted other manufacturers—among them a man named Cleveland E. Dodge Jr.

Clee, as he was known, came from a family of wealthy, connected industrialists. (His grandfather had been a close friend and adviser to President Woodrow Wilson.)[5] For generations, Dodge men had worked in the family's mining concern and moved in New York society circles, but Clee was determined to plot his own course. After graduating from Princeton and serving as a naval commander during World War II, he took a job as a purchasing agent for General Electric. There, he was assigned to a secret program that developed missiles for the fledgling Cold War arms race. However, the project had hit a major roadblock: The wire used to link electronic components kept burning up. While hunting for solutions, Clee came across a Vermont-based firm called Warren Wire. Its products were coated in a revolutionary new plastic called Teflon and were far more durable and heat resistant than wires coated in other materials.[6]

Clee was so excited about Teflon's potential that he quit General Electric to take a job at Warren Wire, settling his family in the Green Mountain foothills. Soon he was trying to convince his new boss to produce other Teflon-coated materials. His boss wasn't interested, so Clee struck out on his own. In 1955, he founded a firm called Dodge Fibers Corporation just across the state border in Hoosick Falls, New York.[7]

Dodge Fibers started out as a modest operation, with only six employees, but before long, it was landing major contracts with aerospace companies. In 1961, Dodge built a mid-sized factory next to the town's waterworks—a location no one thought noteworthy at the time. By the end of the decade, the company had bought several more plants in Hoosick Falls and set up offices in Rhode Island and New Jersey, along with distribution networks in more than a dozen countries.[8]

It was also producing a mind-boggling variety of Teflon-based products—wiring for rockets, insulating tape for electrical work, thread sealant for plumbing, upholstery for cars, and the Teflon-coated aluminum "Slip-foil" that was applied to the bottom of household irons for smoother ironing.[9] Inspired by Dodge's success, other local entrepreneurs got into the Teflon business, too. In 1968, Warren Wire's founder formed the ChemFab Corporation in nearby North Bennington, Vermont. Its specialties included Teflon-infused fabric for use on NASA space missions.[10] Another former Warren employee launched Taconic Plastics, which was based twelve miles south of Hoosick Falls in Petersburgh, New York, and made Teflon-lined conveyor belts for fast-food restaurants, including the wildly popular new chain, McDonald's.[11]

As the companies prospered, so did Hoosick Falls. Shuttered shops reopened; downtown streets were widened to accommodate the surge in traffic.[12] Still, the village held on to its small-town American feel. Sue Hickey remembered the Hoosick Falls of her teenage years as a tight-knit community, where almost everyone turned out for kids' softball games and town band concerts in the park. A slight, blue-eyed girl, Sue had grown up in the Second Ward and went to the local Catholic school. Every day, when classes let out, she would walk through the bustling downtown streets to the supermarket where her father worked as a butcher, browsing the mom-and-pop shop windows and chatting with people along the way. Afterward, she drank ice-cream sodas at Thorpe's Pharmacy or danced the jitterbug with her friends at the Youth Center. Sometimes, on weekends, they drove to the Five Flys nightclub, a former speakeasy that had traded moonshine for go-go dancers and live rock and roll.

During several such outings in the summer of 1968, Sue ran into a garrulous young factory worker with a round face framed by a dark pompadour. His name was John Hickey, although everyone called him Ersel, after a popular rockabilly singer who shared his last name.

The son of a gas station attendant and a factory worker, Ersel had grown up downtown, playing basketball and dreaming about cars. By the time Sue met him, he'd discovered a wild side—he liked to drive fast and occasionally get drunk. The previous summer, Sue saw him plow a convertible into a lake, and came away thinking he was trouble. But he was a good dancer, and he knew how to have fun. They started mixing it up on the dance floor. When Ersel asked her to join him for a Sunday drive through the mountains in his Corvette Stingray, Sue agreed. The following spring, the couple married and moved into a twenty-five-dollar-a-month apartment with a single kerosene heater and a claw-foot tub. Sue's parents lent them a sofa; Ersel, who worked at a factory that made letterpress printing equipment, hocked the Corvette to buy a refrigerator and a stove.

By the mid-1970s, the Hickeys had welcomed their first two children, Katy and Jeff. The letterpress factory had gone under in the meantime, but after frustrating stints as a big rig driver and a railroad worker, Ersel had found a job as a school bus driver. That's when he swore off alcohol—he wanted the parents on his route to feel that he was completely trustworthy.

Bus driving didn't pay much, but Sue worked as a nurse's aide to supplement Ersel's income, and they were able to buy their own home, which Sue (the handy one in the family) slowly fixed up. All the while, the children thrived. Jeff was a natural athlete, and Katy's caring manner endeared her to friends and family. So, in 1978, when the young couple learned they were going to have a third child, they were overjoyed.

Then, when Sue was five months into the pregnancy, she started bleeding heavily. The family's doctor, Philip Martinez, found that her placenta was covering her cervix, a potentially life-threatening condition. Old Doc, as he was known, put her on full bed rest in a hospital in Albany. Even lifting herself onto a bedpan could trigger severe bleeding, doctors there warned. About two months later, Sue started hemorrhaging. The baby, Michael Robert Hickey, had to be delivered eight weeks early.

And so Michael spent the first few weeks of his life in the hospital, with Sue shuttling back and forth by bus from Hoosick Falls to see him. When he was finally released, he weighed under five pounds and had various health problems that would dog him throughout his childhood—asthma, chronic ear infections, and severe allergies, to name a few. Sue eventually gave up her job to care for him, and Ersel took a second job at Dodge's McCaffrey Street factory to fill the gap in income. Every weekday, from 11 p.m. to 7 a.m., he operated a rumbling two-story machine that fed fiberglass cloth through a dip pan full of liquid Teflon and then up a tower lined with ultrahot ovens that fused the Teflon to the fabric. It was his job to man the controls, ensuring the material didn't sag or wrinkle or slide off the rollers.

After punching out from the factory, Ersel drove his morning bus route, then headed home for breakfast and a nap before climbing back into the driver's seat for the afternoon shift. It was a grueling routine, but the factory pay was good enough to lift the family into the middle class—especially once Michael got older and Sue took a job at a plant that made Teflon plumbing tape. Ersel, who loved anything with a motor, bought another Corvette, a Chevy Monte Carlo, and a Yamaha street bike. Even with his demanding work hours, he found time to coach basketball at the Youth Center and hit the courts regularly with Jeff and his buddy Marcus, the youngest of Old Doc's five children.

Throughout these years, Michael often felt like something of an outsider in the family. He was too sickly to play sports like his older brother. (During his one brief foray into basketball in his early teens, he broke both of his wrists, and Sue had to help him bathe and use the toilet for months afterward.) And since Ersel spent most of his waking hours working or coaching, Michael didn't get much time with him. Michael's health problems also took a toll on his self-confidence. He was quieter than his siblings and dreaded confrontation.

Michael didn't find school easy, either. He had a particularly tough time in math and science. But eventually he went away to college, the first person in his family to do so. On summer and winter breaks, Michael stayed with his parents in their new home on McCaffrey Street. The brown-shingled cottage was just across from the original Dodge factory, a spare three-story concrete building now operated by a California-based firm called Furon.[13] When he was in town, Michael worked on the maintenance crew. By then, all the area's factories were drastically scaling up production, mostly to meet demand from Saudi Arabia, which was building a giant tent city out of Teflon-coated fabric to house pilgrims near Mecca. The McCaffrey Street towers operated around the clock, except on Mondays, when they shut down briefly for cleaning.

During the summers, this job fell to Michael. Every Monday morning at five o'clock, he clambered up a coating tower, dismantling the exhaust pipe and scrubbing the thick brown sludge out of it, piece by piece. He'd then haul the parts outside and rinse them in a parking lot near the wells that supplied the village's drinking water. After repeating this process in each of the towers, he'd don a hazmat suit and squeeze into another tall structure that was caked with a dense white crust that had to be ground out with a sander. Then he'd scoop a putrid paste out of the sinks where operators dumped their Teflon dip pans, filling barrels that he carted into the hazardous-materials area. It never even occurred to him to ask what was in the towers—or what happened to the barrels.

After college, Michael lived for a while in Florida before moving back to Hoosick Falls in his late twenties. With that, his life seemed to click into place. He found a good job in the insurance business and started dating a pretty local girl named Angela, who had pale freckles and a mane of curly auburn hair. Ersel was finally getting ready to retire from the factory, and so he and Michael started hanging out almost every day, watching sports and playing golf. Michael cherished the deepening bond with his dad. When Angela got pregnant, Michael imagined his son having the kind of close relationship with Ersel that he'd missed out on as a boy. Now, Ersel's rapid decline was throwing everything into question.

AFTER THE PET SCAN FOUND that the cancer had moved to Ersel's bones, Katy and Michael started spending more time at their parents' house to help care for him. Marcus, who had taken over his father's practice about a decade earlier, stopped by every evening after work and called or texted throughout the day. The treatment options were limited, but he ordered radiation to shrink the tumors and ease the pain. On the morning of Ersel's first appointment, he tried to get up and take a shower but found that he couldn't stand up, even with his walker. Sue had to bring a basin and a washcloth to his bedside. Before long, Ersel's bones were breaking down. Even having a blanket on his legs left him writhing in anguish. Then his white blood cell count plunged. He was transferred to a rehabilitation center in nearby Bennington, Vermont, where the staff could better manage his pain.

The faster Ersel deteriorated, the more determined Michael became to save him. "He was desperate for more time," Katy said. Sometime in February 2013, one of the hospital oncologists pulled Katy aside and suggested it was time to stop treatment and focus on making Ersel's final days more comfortable. When Michael heard this, he angrily questioned the oncologist's qualifications and urged his family to move Ersel to another hospital.

Michael's research took on a fevered quality. He was staying up even later and reading on his phone during family gatherings. On one of his late-night sprees, he came across a novel drug called Votrient, with testimonials recounting seemingly miraculous recoveries from debilitating kidney cancer symptoms. Ersel's insurance agreed to cover the $8,500 monthly cost. And almost as soon as he began the regimen, Ersel seemed to improve—sitting up in his bed, he started chatting about the high school basketball season. Sue was so relieved that she didn't even protest when he mentioned buying a new RV.

The reprieve lasted only a few days before Ersel's heart started beating erratically, a common side effect of cancer treatments. He was rushed to the hospital and jolted with a defibrillator. Afterward, Marcus met the family in the emergency room and advised them to abandon treatment. Michael bristled: He was convinced that the

Votrient could still work. But Sue had no interest in prolonging her husband's suffering. She instructed the hospital to send Ersel home and dispatched Michael and Jeff to clear out the ground-floor bedroom for a hospital bed.

From that point on, the entire Hickey family camped out at the house. Michael changed Ersel's soiled undergarments. Katy moved fentanyl patches around his back and doled out sedatives, carefully logging times and doses in a notebook. Two-year-old Oliver dragged Ersel's favorite Cleveland Browns blanket and an old rocking chair into the bedroom to keep his "pops" company. As for Ersel, he mostly slept. When he was awake, his brain struggled to make sense of what was happening to his body. Sometimes he wailed in agony. Michael was at his bedside one morning when Ersel turned to him and gasped, "I'm fucking dying! Somebody's got to help me."

By then, Ersel's body was shutting down. His nail beds grayed. His mouth gaped and his breathing reverberated with an eerie crackle. As the sun sank behind the soft green hills on the evening of February 25, Marcus and the family gathered at his bedside. Michael pulled off the oxygen mask and started squeezing morphine into Ersel's mouth, but his hands were too shaky; Marcus had to take over.

The medicine was supposed to ease Ersel's passage, but Ersel wasn't ready to go. He howled and thrashed and retched with every last scintilla of his strength. Then Sue started howling, too, and had to be carried out of the room. Eventually, Michael got behind the bed and held his father's bony shoulders down to keep him from flailing. Ersel struggled for another ten or fifteen minutes before finally going limp. Michael, Jeff, and their uncle Rich stripped the soiled sheets off his bed and washed his body. Then Michael drove to a local dive bar and started pounding shots of whiskey.

A few days later, as the family was preparing for the wake, Michael climbed into his parents' attic to look for pictures. The room was crammed with squat plastic barrels, some filled with

Christmas decorations and winter clothes. For as long as Michael could remember, workers at the local Teflon factories had brought these containers home and used them for everything, from gardening to bobbing for apples. But this time he noticed something ominous that he hadn't seen before. On one barrel, he could still make out some words on the faded original label: a warning about "toxic gases."

3

Lucifer's Gas

On the morning of April 6, 1938, a young chemist named Roy Plunkett arrived at a DuPont plant in Deepwater, New Jersey, and dispatched his assistant to fetch a special canister from the dry-ice chamber.[1] The two men were on a nerve-racking quest to develop a new refrigerant based on one of the most feared substances known to science, the element fluorine. Sometimes called Lucifer's gas, fluorine burst into flames in the presence of water and burned through virtually every other material—glass, wood, concrete, even steel. Most of the scientists who had dared to work with it over the past two centuries had wound up injured or dead.[2] But one of Plunkett's colleagues had managed to transform the pale-yellow gas into the refrigerant Freon, which was being produced exclusively for Frigidaire.[3]

Plunkett's charge was to develop a similar product that could be marketed more widely. As a first step, he'd chosen to make a substance called tetrafluoroethylene, or TFE, which had been synthetized only in tiny quantities, and to produce the previously unimaginable volume of one hundred pounds.[4]

Plunkett's assistant returned carrying a fat black canister containing the first fruits of their efforts. The pair attached it to a testing apparatus and swiveled the valve, expecting TFE gas to billow forth. Nothing happened. Plunkett fiddled with the nozzle, then removed the valve and shook the canister over the bench. A flaky white powder drifted out. Fearing the experiment had failed, he began sawing the other canisters open and found them lined with the same waxy white substance.[5]

Plunkett was flummoxed. He suspected that the TFE had polymerized—that its molecules had arranged themselves into long, repeating chains typical of synthetic plastics. But the material had peculiar properties that other plastics didn't. It was extremely slippery. It refused to melt, even when he zapped it with a soldering iron and an electric arc. Plunkett tried dissolving the stuff in acetone, ether, alcohol, hot water, pyridine, nitrobenzene, sodium hydroxide, and concentrated sulfuric acid. Nothing broke it down. "It didn't react with anything," Plunkett later recounted. "Those facts themselves said there was something unusual about this material."[6]

Plunkett's accidental invention—which is now known by its brand name, Teflon—would become revolutionary both for DuPont, which parlayed it into a billion-dollar-a-year business, and for American industry. But the discovery would have gone nowhere without the U.S. government, which spent huge sums transforming Teflon from a laboratory oddity into a viable product. The investment was part of a centuries-long partnership between the government and DuPont—a relationship that was crucial to the rise of the U.S. chemical industry and the synthetic materials that are now so integral to our daily lives.

That alliance stretched all the way back to one of the nation's founders. When Thomas Jefferson was serving as a diplomat in France in the late 1700s, he befriended the French nobleman and philosopher Pierre Samuel du Pont de Nemours.[7] The pair struck up a lively correspondence, which continued after Jefferson's return to

America. ("It is rare I can indulge myself in the luxury of philosophy," Jefferson wrote. "Your letters give me a few of those delicious moments.")[8] After the French Revolution, du Pont de Nemours, who had helped to physically defend Marie Antoinette and King Louis XVI from a marauding mob, was forced into hiding. He later fled with his family to the United States, but storms slowed their ship's crossing. During the three months they spent at sea, the family survived on a stew made of boiled rats. By the time they reached Rhode Island in early 1800, they were literally starving.[9]

It didn't take long for them to rebound from these grim circumstances, though. In 1803, Pierre helped his old friend Jefferson—who by then was president of the United States—negotiate the Louisiana Purchase. Meanwhile, Pierre's son built a gunpowder mill along Delaware's Brandywine River.[10] In a letter to Jefferson, the younger du Pont de Nemours expressed his hope that the president would "look favorably" on the factory and grant it munitions contracts. Jefferson obliged, citing his "debt of gratitude" to Pierre. In July 1803, he urged his secretary of war to buy gunpowder from the budding enterprise. This patronage helped establish E.I. DuPont de Nemours & Company as the government's chief munitions supplier for decades to come.[11]

By the late 1800s, DuPont was also the country's largest supplier of nonmilitary munitions. And it even held sway over the powder mills it didn't own outright, thanks to the Gunpowder Trade Association, a cartel of ostensibly rival companies that carved up markets and fixed prices.[12] DuPont's military contracts alone during this era netted up to $3 million per year (roughly $104 million in today's dollars), mostly from the sale of a particularly effective form of smokeless powder that didn't envelop the battlefield in a blinding haze. Although the U.S. Navy had invented the material, DuPont had exclusive rights to produce it.[13]

Yet one member of the DuPont clan—a shrewd, gawky MIT-educated chemist named Pierre S. DuPont—was convinced the company could become even more profitable. In 1902, he and two cousins bought out the rest of the family. The DuPont company then proceeded to gobble up most of its competitors, including former cartel members, and organize the hodgepodge of factories and products into a centrally managed corporation. Pierre's team also revamped the financial operations, developing a groundbreaking formula that combined earnings, working capital, and other assets into a single measure: *return on investment*. This approach—which has become a defining principle of global business—left little room for the considerations, like family loyalty, that had once shaped DuPont's operations. Profit was the only yardstick that mattered.[14]

Meanwhile, President Theodore Roosevelt took office and, seizing on the recently passed Sherman Antitrust Act, launched his trust-busting campaign.[15] This handed DuPont's few remaining competitors a potent weapon. In 1906, a former DuPont sales manager named Robert S. Waddell, who had formed his own powder company, began bombarding Congress with letters and testimony.[16] In them, he accused DuPont of using "every vicious and disreputable act known in the catalog of trust warfare," from price-fixing to corporate espionage. Such tactics, Waddell argued, had given the company a near-total monopoly on the powder business, allowing it to bilk the government out of millions of dollars each year. He also alleged that DuPont's dominance was a threat to national security, since the entire munitions supply was "hopelessly at the mercy" of a single firm.[17]

Waddell's claims ricocheted through the media. After dispatching an investigator to Peoria to peruse his papers, the Justice Department charged DuPont with various antitrust violations; Congress began weighing legislation to bar the military from buying the company's munition. DuPont fought back with a lobbying offensive targeting lawmakers and DuPont's allies inside the military.[18] It also took a step that would have seismic implications for American industry as a whole: launching its own scientific research operation. At the time, very few U.S. firms had their own R&D labs; most groundbreaking innovations came from independent inventors. But DuPont was determined to monopolize scientific understanding of nitrocellulose, the main ingredient in smokeless powder, which was crucial to America's arsenal. Executives believed doing so would allow DuPont to stay "well in front of the military" in terms of technical expertise, ensuring the government would remain dependent on the company.[19]

When the court delivered its verdict in 1912, it found that both DuPont and Pierre personally had violated antitrust laws and ordered the company to spin off part of its dynamite and gunpowder business.[20] On the surface, this looked like a win for the trustbusters, but in reality DuPont surrendered only about a quarter of its assets. And it got to hold on to its entire smokeless powder operation—the ostensible reason the case was brought in the first place. This was a concession to army and navy ordinance chiefs who, at the company's urging, had testified to their reliance on DuPont's production capacity and scientific expertise. Echoing their arguments, the court found that breaking up its smokeless powder holdings could cause "injury to the public interests of a grave character."[21]

Two years later, when World War I engulfed Europe, demand for munitions skyrocketed. Thanks largely to its lock on the smokeless powder market, DuPont's annual profits soared to more than ten times their average in the decade before the war.[22] But the conflict also brought dire new threats to the company's business. At the time, the Germans held a near-total monopoly on the development and production of synthetic chemicals. The Allied blockade of Germany drastically reduced America's supply of the chemical ingredients for gunpowder and caused nationwide shortages of synthetic materials. So DuPont hired a team of chemists to reverse-engineer substances like toluene that it needed for its production lines. Soon, it was churning out tens of thousands of pounds of these chemicals each month. More important, it had begun to grasp what one internal memo described as the "common scientific basis" underlying these seemingly diverse compounds, all of them derived from a few fossil-fuel-based industrial by-products, such as coal tar.[23]

The deputy director of DuPont's development department recognized the enormous business potential of conjuring marvelous products from the dregs of other industrial processes. He pressed the executive committee to build a chemistry operation rivaling the one the Germans had cultivated over decades, arguing it would pave the way for a vibrant "system of new industries." The committee agreed. And in 1917, DuPont broke ground on a vast laboratory complex in Deepwater, New Jersey, which would soon employ some 550 people.[24]

DuPont also began buying up the few U.S. companies that produced synthetic materials and aggressively pursuing German technical know-how. It recruited U.S.-based employees of German chemical makers and lobbied the American government to hand over the patents seized from the German firms' U.S. operations.[25] The Office of Alien Property Custodian, which oversaw seized German assets, eventually gave DuPont and another firm access to the German documents. Lawyers for the companies worked in shifts around the clock, creating lists of patents to be transferred to U.S. industry.[26]

After the war, forty-five hundred patents worth tens of millions of dollars were sold to the Chemical Foundation Institute, a new nonprofit underwritten primarily by DuPont, for a mere $250,000.[27] The founding alien property custodian, A. Mitchell Palmer, believed that putting German expertise in the hands of American companies was vital to national security. Germany had channeled its command of chemistry into vicious new weapons, including poison gases, and punished Allied economies by withholding raw materials. The best way to defend against such threats in the future, Palmer argued, was to build up America's own chemical industry. "The next war, if it ever does happen, will be a chemists' war, and the country which has the best-developed dye and chemical industry is the country which is going to come out on top," he wrote in 1920.[28]

As it turned out, the patents alone were often useless: To foil would-be competitors, German chemical interests had intentionally made many of the documents vague or misleading. So DuPont launched a covert mission to recruit German chemists who could interpret them. A Swiss-born DuPont scientist named Eric Kunz was dispatched to Europe to hunt for candidates. Most of those he approached were wary of aiding the Americans. But a handful traveled to Switzerland for interviews with DuPont executives, and four PhD chemists signed generous employment contracts.[29]

In late 1920, Kunz spirited several recruits to the Netherlands. Along the way, Dutch border guards seized a trunk stuffed with technical drawings and alerted German authorities, who issued arrest warrants for the chemists on charges of industrial espionage. Two of them managed to board a Dutch steamer bound for New York. The ship's captain tried to detain them at Ellis Island, citing the warrants, but a DuPont official posing as president of an international police order met him at the port and negotiated their release.[30] At DuPont's request, the commander of U.S. forces in Germany later directed the military police to track down the two remaining chemists, and in July 1921, the pair arrived in New Jersey aboard an American military transport.[31]

FLUSH WITH WARTIME PROFITS, DuPont began a major expansion based largely on the mass production of German innovations. The firm also staged a hostile takeover of the upstart automaker General Motors, aiming to become the sole supplier of synthetic materials like paints and plastics for GM's assembly lines. Between 1917 and 1919, DuPont, and Pierre personally, invested about $50 million in GM, making them the largest stockholders. Pierre also helped to oust the company's founder, William C. Durant—after which

Pierre became GM's president, while his brother took the helm at DuPont.[32]

Pierre quickly set to work modernizing the automaker's operations. Under his leadership, every decision would be based on DuPont's novel metric: return on investment.[33] This approach extended to the company's research labs in Dayton, Ohio, where scientists and engineers had focused on blue-sky projects. Now they came under intense pressure to deliver profit-making innovations—especially ones that could lead to major production contracts with DuPont. Thomas Midgley, an enterprising young poet and chemical engineer, was assigned to develop a fuel additive that would eliminate the knocking then common in car engines. This would improve horsepower and fuel efficiency—a major selling point at a time when the domestic oil supply was believed to be on the brink of depletion.[34]

Midgley worked his way through thousands of substances, from aluminum chloride to melted butter. Most of them, Midgley later recalled, "had no more effect than spitting in the Great Lakes." But he eventually found two that solved the problem. One was ethanol, or grain alcohol, which is primarily made from excess crops and the stalks and leaves left over after harvest. The other was a toxic compound called tetraethyl lead. In a 1922 letter to his brother, Pierre DuPont described the substance as "very poisonous if absorbed through the skin, resulting in lead poisoning almost immediately." But unlike ethanol, the tetraethyl blend could be patented—meaning GM and DuPont would profit on every gallon. GM opted for the leaded formula.[35]

By the summer of 1924, DuPont was manufacturing the additive at its new tetraethyl blending plant in Deepwater, New Jersey, part of an exclusive deal with GM. And plant workers were being hauled away to a hospital, twitching, thrashing, and howling at imaginary phantoms—the physical and psychiatric symptoms that often accompany acute lead poisoning. At least eight died, some of them in straitjackets.[36] DuPont, which owned some of the region's largest newspapers, managed to keep the deaths quiet.[37] But that August, GM and Standard Oil began testing a new method for blending leaded gasoline at a Standard plant near Elizabeth, New Jersey. Soon after, one of the workers began tearing around the plant in terror, screaming, "Three coming at me at once!" Another man hurled himself out of a second-floor window. A third worker went home sick and woke up the next morning raving and violent. It took four men just to get him into a straitjacket.[38]

By late October, five of the plant's forty-nine employees were dead; thirty-five more had been hospitalized. This time the carnage was national news. ("Many Near Death as Result of Inhaling 'Looney Gas' Fumes," read a headline in one Nebraska paper.) New York City, Philadelphia, and parts of New Jersey halted the sale of leaded gasoline, which was already being pumped at about twenty thousand stations around the country.[39] Scientists called for a nationwide ban until its effects could be studied. "This is probably the greatest single question in the field of public health," one prominent Yale physiologist told a gathering of engineers. "It is the question whether scientific experts are to be consulted, and the action of the government guided by their advice, or whether, on the contrary, commercial interests are to be allowed to subordinate every other consideration to that of profit."[40]

Desperate to contain the damage, in late 1924 executives from DuPont, Standard, and GM paid a private visit to the U.S. surgeon general. Seeking to dissuade him from ordering government studies on leaded gasoline, they urged him to convene a conference where industry, scientists, and public health officials could determine the best path forward.[41] DuPont and GM also hired scientists to help them frame the debate on leaded gasoline. Chief among them was a young toxicologist and University of Cincinnati professor named Robert Kehoe.

A short, dark-haired man with formidable eyebrows, Kehoe had a quasi-religious faith in the power of technological progress to solve society's problems. Any downsides, he would later argue, were simply "the prices which must be paid for the privilege (and the necessity) of living in a technological era." They were far outweighed by the benefits that scientific advances bestowed on humanity and could usually be managed using "appropriate technological means."[42]

In the case of leaded gasoline, Kehoe discovered that the poisonings at the plants were due to dangerous levels of tetraethyl lead accumulating on the factory floor—a problem that could be fixed with simple changes like hip boots and fans to disperse the fumes.[43] He also performed a study comparing the levels of lead in mechanics, gas station attendants, and tanker-truck drivers who handled leaded gasoline with people from mostly similar backgrounds who supposedly hadn't been exposed.* Both groups harbored lead in their bodies, and they scored roughly the same on various health indicators. Based on these results, Kehoe concluded that leaded gasoline posed no threat to the general public—and that low levels of lead occurred naturally in the human body.[44]

The surgeon general eventually convened the conference that the industry had requested. Some of the prominent scientists in attendance raised concerns that the public could be gradually poisoned by lead fumes over time.[45] Although Kehoe's lone short-term study couldn't rule this out, he brushed aside their concerns with his usual swagger, arguing there was no point in banning useful products unless there was hard proof of "actual danger to the public."[46] In the end, Kehoe prevailed. The surgeon general appointed a committee to conduct a brief investigation, which found "no good grounds" for banning leaded gasoline. The group also recommended publicly funded studies to assess its long-term health effects. But Kehoe and

* It would later come to light that Kehoe's control group included garage workers who didn't handle lead directly but were nonetheless saturated with lead from tailpipes and other sources.

industry leaders persuaded health officials to let them handle this work, arguing that companies, not taxpayers, should shoulder the considerable expense.[47]

These events opened the door to the widespread sale of leaded gasoline, which would wreak havoc on public health for decades. They also established the bedrock principles that have governed our system for regulating potentially harmful substances ever since: first, that industry can be trusted to serve as an unbiased arbiter of science, and second, that products are to be presumed safe until proven otherwise—a premise now known in public health circles as the Kehoe Principle.[48] This approach created a powerful incentive for corporations to gin up doubt about research linking their products to disease. As long as questions remained about the "actual danger to the public," there would be no regulation.

Kehoe himself would emerge as the chief purveyor of scientific doubt. In 1929, DuPont, GM, and their subsidiaries gave the University of Cincinnati medical school a generous sum to create a toxicology research program directed by Kehoe.[49] This entity became a leading source of information on the health effects of industrial pollution and consumer goods. Because virtually all its funding flowed from corporations—and contracts specified that it couldn't publish data without funders' approval—manufacturers had enormous power to shape the science and public perceptions about the safety of their products.[50]

Innovations like leaded gasoline transformed both GM and DuPont. By the late 1920s, GM had become one of the world's largest companies, with the highest annual profits reported by any industrial firm before World War II.[51] This translated into huge dividends for DuPont. The head of the company's chemical division, Charles M. A. Stine, decided to use the windfall to test the boundaries of his discipline.[52]

A preacher's son, Stine had long viewed science as a means of understanding God's designs. But after World War I, he came to see it as something even more transcendent. Through chemistry, he argued in a speech before the American Chemical Society, God had given man the power not only to "emulate Nature" but also to "excel her in certain fields of creation."[53]

To prove his vision, in 1926 Stine proposed launching a "pure research" program to investigate the basic properties of chemicals instead of merely using established science to solve specific practical problems. This approach, he argued, would lend DuPont prestige, making it easier to recruit PhD scientists. It could also lead to radical breakthroughs that would benefit humanity—while opening vast new markets.[54]

DuPont's executive committee gave its approval, and Stine began recruiting, though he had trouble finding accomplished scientists who wanted to work for industry. After a frustrating nine-month search, he approached a brilliant young Harvard instructor named Wallace Carothers about heading the synthetic chemistry division. Carothers relished the idea of a research career free of teaching obligations, but he worried that he would be pressured to do work with profit-making potential. In addition, as he confided in a letter to one of Stine's colleagues, he suffered from "neurotic spells of diminished capacity which might constitute a much more serious handicap" at DuPont than in academia. Stine responded by raising the already-generous salary offer and dispatching a deputy to Cambridge with orders not to leave until Carothers accepted.[55]

After arriving at DuPont in early 1928, Carothers threw himself into researching polymers, small molecules linked together in long, repeating chains to form much larger molecules.[56] Scientists had long known that certain natural materials—including silk, wood, and DNA, the basic building block of life—were structured this way. In a few cases, researchers had stumbled on synthetic polymers with fabulous properties. Bakelite, for instance, resisted electrical currents and could be molded into thousands of objects, from jewelry and toys to machine parts.[57] But scientists were divided over how to produce these materials, commonly known as plastics. The prevailing theory was that polymers were bound by completely different forces than smaller molecules. Recently, however, a prominent German chemist had uncovered evidence that they were held together by ordinary molecular bonds. "We were as shocked as zoologists might be if they were told that somewhere in Africa an elephant was found who was 1,500 feet long and 300 feet high," one of his colleagues remarked at the time.[58]

Carothers set out to prove this theory. Sitting in DuPont's third-floor library, overlooking the tulip poplars along the Brandywine, he sketched out methods for stringing atoms into giant molecules using ordinary chemical reactions. This work—which he later distilled into two landmark papers—would help lay the foundation for the modern plastics revolution.[59]

The first signs of this revolution surfaced on April 17, 1930, when one of Carothers's lab assistants found a clear solid mass in a test tube he had left sitting on his research bench. It turned out to be a new and miraculously versatile form of synthetic rubber. Ten days later, the lab made another breakthrough. Carothers's quest to build giant molecules had run into a wall when they passed a certain size. Suspecting the problem was water formed during the reaction, he tweaked a piece of lab equipment to draw off moisture. The resulting material had more than double the molecular weight of anything his lab had produced. When his colleague dipped a glass rod into a heated mass of the substance, he pulled out a "festoon" of gossamer fibers that could be stretched to several times their original length. And the longer they grew, the stronger and more elastic they became. The lab had invented the world's first fully synthetic fiber.[60]

These materials were scientific marvels, but the early versions had limited commercial appeal. The fiber melted at too low a temperature to be used in clothing. The synthetic rubber—variations of which were later marketed as neoprene—was less durable and more costly than natural rubber. Plus, it gave off a horrid odor. With the Great Depression eroding DuPont's profits, Carothers came under growing pressure to come up with lucrative innovations. As he'd feared, his black moods resurfaced with a vengeance. In a 1932 letter, he described himself feeling "feeble, smelly and cockroach-like. Just why, I don't know. At any rate I go through at least a dozen violent storms of despair every day."[61]

Hoping to steer him in a more productive direction, DuPont's new director of chemical research encouraged Carothers to revisit his work on synthetic fiber. Sure enough, by 1935 Carothers's lab had developed a substance that had all the strength and elasticity of his original creation but could withstand heat and moisture well enough to be used in clothing. DuPont resolved to mass-produce the material, known as Nylon 66. But Carothers was too consumed with depression to celebrate. Just after filing the patent application for nylon, he checked himself into a Philadelphia hotel and swallowed a fatal cocktail of lemon juice and cyanide.[62]

BY THEN, DuPont had become a household name—though not for its scientific discoveries. In 1934, a journalist and historian had copublished a blockbuster book about the U.S. arms industry with the incendiary title *Merchants of Death*. It argued that DuPont had unduly influenced America's decision to enter World War I, then reaped exorbitant profits by supplying America's enemies as well as Allied forces.[63] Congressional investigators spent three days grilling Pierre and another company executive about the accusations.[64] A separate congressional probe uncovered a bizarre plot—allegedly funded by DuPont and other companies that opposed the New Deal—to overthrow the U.S. government and install a Mussolini-style dictatorship.[65] Almost overnight, DuPont became a national pariah, and Congress threatened to cancel its munitions contracts.

Desperate to salvage its reputation, the company hired a legendary PR consultant named Bruce Barton. Barton concluded there was only one way DuPont could escape the "atmosphere of plague": It had to transform itself in the public's mind from a maker of deadly munitions to a source of marvelous inventions that benefited society.[66]

So DuPont introduced a new slogan: "Better Things for Better Living . . . Through Chemistry." It wasn't so much a catchphrase as a blueprint for a utopian future, one where mankind would harness the wonders of science to overcome the limitations of nature and create a safer, more beautiful world. In 1938, Charles Stine personally unveiled the first of these revolutionary materials at the site of the upcoming New York World's Fair. The fabric was called nylon, he declared. Made from "coal, water and air," it could be fashioned into fibers "as strong as steel, as fine as the spider's web." DuPont later opened a wildly popular fair exhibit called Wonder World of Chemistry, which featured a shapely Miss Chemistry rising out of a test tube in a nylon evening gown and stockings, alongside a hundred-foot tower that resembled a laboratory apparatus.[67] These fantastical visions, amplified by exuberant media coverage, captured the American imagination. When the new nylon stockings went on sale in 1940, they sold out almost immediately.[68]

The same year that Stine introduced the world to nylon, Roy Plunkett accidentally invented Teflon. DuPont immediately recognized its enormous potential. Besides its uncanny strength and heat resistance, it repelled water and grease and stood up to chemicals that ate through most other substances. One of DuPont's first tests involved a dye plant using chemicals so corrosive that the pumps reportedly had to be replaced almost weekly. When the pumps were packed with Teflon, they could operate for months.[69]

But transforming this promise into actual products proved remarkably difficult. DuPont's polymer chemistry division, by then the most advanced in the world, couldn't find a way to manufacture Teflon at scale. And the same qualities that made the material so potentially useful also made it incredibly challenging to work with. It was extremely slippery. It didn't melt or dissolve in other chemicals, so it couldn't be molded using the usual methods. After taking stock of these obstacles, DuPont's executives concluded the substance would probably never be put into production.[70] What they didn't yet know was that their stubborn creation was about get a boost from an unexpected quarter—a vast, secret program run by the U.S. government.

4

Exile to Devil's Island

On a clear, hot afternoon in July 1939, a car carrying the esteemed physicists Leo Szilard and Eugene Wigner rumbled up the drive of a humble white cottage on Long Island's Cutchogue Harbor, the verdant retreat where Albert Einstein was passing his summer.[1] In recent years, scientists around the world had begun experimenting with splitting uranium atoms, a process with the potential to release virtually unlimited energy. Some believed it could also lead to ferocious new weapons. Wigner and Szilard, both Hungarian-born Jews who had studied under Einstein in Berlin before the rise of Adolf Hitler, were convinced that the Nazis were already on the brink of making an atomic bomb. But their desperate efforts to warn U.S. officials had led nowhere, so they decided to approach the one person in their orbit who they believed could get their message through.[2]

Dressed in an undershirt and rolled-up pants, Einstein greeted his old friends and led them to a screened-in porch overlooking a sloping lawn. There, between sips of iced tea, Szilard and Wigner explained the scientific reasons for their fears. In the past, Einstein had voiced skepticism about the chances of harvesting large-scale nuclear energy in the near term. But Wigner and Szilard persuaded him it was possible. Although he was an ardent pacifist, Einstein agreed to help alert the U.S. government to the threat, and on August 2, 1939, a spare two-page letter to President Franklin D. Roosevelt went out under the renowned physicist's name. It urged the president to speed up experimental work on uranium by underwriting university research and "obtaining the co-operation of industrial laboratories which have the necessary equipment."[3]

Heeding this warning, the Roosevelt administration began secretly funding research by a team of physicists at Columbia University. The goal was to isolate a rare class of uranium atoms that were capable of producing nuclear chain reactions—the only process that could yield enough energy for an atom bomb. Separating the minuscule particles from the rest of the uranium ore presented mind-bending technical challenges. But the physicists devised several possible methods. The most promising—gaseous diffusion—involved converting uranium into a gas called uranium hexafluoride, or hex, and pumping it through a maze of porous barriers. Since the desired isotope, uranium-235, passed through the tiny pores more easily, the rest of the atoms would gradually be filtered out, leaving only the prized nuclear fuel.[4]

However, because hex also contained fluorine, the infamous Lucifer's gas, the compound was dangerous to work with and fiercely corrosive, making it extremely difficult to contain.[5] If the project stood any chance of succeeding, the physicists needed materials that could stand up to both fluorine and hex in some of the harshest conditions imaginable.

The Nobel Prize–winning scientist overseeing the uranium program, Harold Urey, suspected that only other compounds containing fluorine would work—specifically those containing fluorine and carbon, which together form the strongest bond in chemistry. Such materials were extremely rare, but Urey managed to track down a few drops of a fluorocarbon liquid, the result of a laboratory accident at Pennsylvania State University. Sure enough, when mixed with the hex, it produced no reaction.[6]

This was a major breakthrough. But making enough uranium for a bomb would require mind-boggling quantities of hex-resistant equipment. Urey's team would need an assortment of fluorocarbons suited to various purposes. Among the most urgent was a fluorocarbon plastic that could be fashioned into seals and gaskets to keep the enrichment system airtight. As luck would have it, DuPont had already developed one that seemed to fit the bill: Teflon. The company hadn't figured out how to make more than a few ounces at a time, but it had a history of ramping up production fast.

Urey's deputy eventually summoned Malcolm Renfrew, the chemist overseeing DuPont's stalled Teflon program, to Columbia, by then the bustling hub of the bomb project. Without revealing the exact nature of the enterprise, he pleaded for Renfrew's help. "He told us there was a development now coming on in this country and in Germany which would determine who would win the war, that it was going to be extraordinarily important for us to be participating at our maximum strength," the chemist recalled. Renfrew would be given a few weeks to prepare, and then DuPont was expected to break ground on a plant that manufactured a million-plus pounds of Teflon per year.[7]

The conversation left Renfrew "popeyed," as he later recalled. Nevertheless, DuPont agreed to launch a government-funded crash program to figure out how to produce Teflon at scale. A team of chemists and engineers at Renfrew's lab in Kearny, New Jersey, worked around the clock building a pilot plant. Almost as soon as construction was finished, the plant exploded, killing two young workers. The crew rebuilt it—this time with blast walls and remote controls to handle the most dangerous work.[8] Meanwhile, another DuPont team studying molding techniques discovered that a combination of pressure and extreme heat melded Teflon powder into sheets that could be sliced and sculpted into hex-proof gaskets.[9]

By late 1941, the government was recruiting chemists from venerable institutions like Purdue and Cornell to cultivate other types of fluorocarbons—among them refrigerants, sealants, and lubricants—and overcome the daunting technical barriers to large-scale production.[10] A team from Johns Hopkins worked with DuPont to develop industrial methods for isolating fluorine and manufacturing other previously scarce substances essential for fluorocarbon production.[11] The university scientists weren't told why their expertise was needed, and the coordination among them was initially haphazard.[12]

That all changed in July 1942. Urey convened a secret meeting with military officials, DuPont executives, and chemists from various universities at Dumbarton Oaks, an august estate in Washington's Georgetown neighborhood. There, he announced plans to "sponsor closer collaboration" among attendees. The primary goal was to develop a specific class of fluorocarbons whose defining feature was multiple fluorine-carbon bonds, making them virtually indestructible—a group of substances that would later become known as forever chemicals or PFAS. In the interest of speed, participants were told, the government was willing to fund all lines of inquiry simultaneously and pay generously for the resulting materials. ("Dollar cost will be a very small factor," the meeting minutes noted.)[13]

The event touched off a technological race among project chemists that mirrored the urgent, all-at-once approach to developing nuclear fuels. Because of DuPont's unrivaled experience with fluorine compounds, it was charged with coordinating their efforts. Chemists at various universities compiled their data into monthly reports for DuPont's research director, who kept Urey updated and helped rush discoveries into production.[14] In the last two months of 1942 alone, the government contracted with DuPont to build two factories to produce fluorocarbon lubricants and sealants based on research from university scientists, and a third facility to manufacture a chemical critical to the production of both fluorocarbons and hex—which were now a matter of national security.[15]

IT WAS AROUND THIS TIME that the Roosevelt administration tapped army general Leslie Groves to direct the secret bomb program. A portly, plainspoken engineer, Groves was charged with organizing the scattered research on nuclear fuel and managing the transition to warhead production. By this point, Columbia's gaseous diffusion program had spilled over into a former car dealership in Harlem. Physicists at the University of California, Berkeley, were working on an alternative method for enriching uranium.[16] A third group at the University of Chicago was building the world's first nuclear reactor on an abandoned squash court under the west stands of Stagg Field, hoping to manufacture another promising nuclear fuel: plutonium. So far no one had managed to produce more than a few specks of this substance, and when the Chicago team fired up the reactor, there was a chance it would veer out of control, triggering a full-blown nuclear explosion.[17]

Still, Groves determined that the plutonium process was most likely to succeed. To produce the necessary volume on a breakneck timeline, though, he needed a contractor that could tackle engineering, construction, and manufacturing on a mammoth scale. "Only one firm was capable of handling all three phases of the job," Groves concluded. "That firm was DuPont."[18]

In October 1942, Groves arranged a meeting in Washington, D.C., with Charles Stine, who had overseen the early development of nylon and was now a member of DuPont's powerful executive committee. Up until this point, DuPont had been given only as much information about the bomb project as it needed for discrete assignments. Now, Groves revealed the program's staggering scope and explained that he wanted DuPont to handle a critical component: plutonium manufacturing.[19]

Stine and the chemist who accompanied him found the proposal bewildering. According to Groves, they "protested vigorously that DuPont was experienced in chemistry not physics, had no knowledge or experience in this field, and that they were incompetent to render any opinion except that the entire project seemed beyond human capability."[20]

In fact, no institution on earth had ever attempted anything remotely similar to what Groves was proposing. Groves himself wasn't sure it was possible. But he *was sure* the program stood a better chance of success if it could tap into DuPont's considerable production prowess. So he dialed up the pressure. In a series of secret meetings with DuPont's president and executive committee, Groves emphasized that the project was of the "utmost national urgency," that the Germans were likely working on an atom bomb, and that "there was no known defense against the military use of nuclear weapons except the fear of their counter-employment."[21]

The rest of the executive team shared Stine's skepticism, and it wasn't only the stunning complexity of the task that worried them. DuPont was still struggling to shed its reputation as a "merchant of death."[22] Just a few months earlier, Congress had learned that a GM joint venture had cut a deal in the late 1930s to supply the Germans with the secret for making leaded gasoline, without which Hitler could not have gone to war. Lawmakers were livid.[23] DuPont's leaders feared their company would be seen as profiting from the devastating new atomic weapons and that the project would open it up to enormous liability.

To allay these concerns, that fall Groves invited Stine back to Washington for a meeting with him and James B. Conant, then president of Harvard and one of the bomb project's chief scientific advisers. The pair offered to allow a delegation made up primarily of DuPont engineers to visit the three nuclear fuel programs and assess for themselves which of the methods would be most feasible to put into mass production.[24]

Stine accepted the invitation. And in November 1942, a committee made up of three DuPont engineers and chaired by the MIT chemist Warren K. Lewis traveled by train from New York to Chicago and then Berkeley, inspecting equipment and grilling Nobel Prize–winning physicists as they went.[25] "We acted like we were the Supreme Court," one of the DuPont men recalled. "We just tormented the life out of those people, asking questions." Between stops, they huddled in their double compartment, reading secret reports, jotting notes in loose-leaf notebooks, and sipping scotch or ginger ale.[26]

The Lewis Committee happened to arrive back in Chicago just as the team there was finishing work on its reactor. The physicists were all too aware that the DuPont engineers took a pessimistic view of the project. Arthur Compton, the Nobel laureate overseeing the plutonium work, had warned in a letter to Conant and Groves that the committee could relegate the project to "so low a status that it could not be of value in this war."[27]

Hoping to change the committee's view, Compton invited one of the DuPont delegates, Crawford Greenewalt, to be present when they started the reactor—a crucial test for the plutonium program. On the afternoon of December 2, 1942, Greenewalt joined the University of Chicago scientists on the balcony overlooking the squash court as the eminent physicist Enrico Fermi ordered the last control rod pulled out of the reactor. The neutron counter began clattering, first slowly, then faster and faster, until the clicks merged into a single steady hum. Fermi thrust his hand in the air to signal that the reactor had gone critical, triggering the world's first nuclear chain reaction.[28] In his diary, Greenewalt called the event "thrilling" to witness and the outcome "much better than expected."[29]

Yet when the committee delivered its report five days later, the authors reiterated their doubts about the plutonium process. They recommended pursuing all the approaches but devoting the lion's share of resources to producing uranium fuel via gaseous diffusion, which they argued had the "best over-all chance of success." Notably, this was the option that top bomb-project officials had previously considered the *lowest priority*.[30] It was also the one that required large quantities of Teflon, making it potentially lucrative for DuPont.

The government body overseeing bomb research endorsed the Lewis Committee's road map in a letter to President Roosevelt.* And on December 8, 1942, the president authorized formation of the Manhattan Project, which gathered the various entities working on the bomb into a sprawling complex that would cost billions of dollars and employ more than a hundred thousand people—all in total secrecy.[31]

Meanwhile, DuPont signed on as the bomb project's main plutonium producer, provoking a near rebellion among the Chicago scientists; one Nobel Prize–winning physicist threatened to quit.[32] To avoid being seen as a war profiteer, the firm agreed to limit its fee for the plutonium project to one dollar above costs and to turn all plutonium patents over to the U.S. government. But it held on to its patent for Teflon, expecting the material would be key to the Columbia method for enriching uranium.[33]

In keeping with the Lewis Committee's recommendation, this method was now the top priority. While most of the fuel programs were relegated to the pilot-plant phase, work immediately began on a full-scale gaseous diffusion plant.[34] The scope of the operation was breathtaking. Composed of fifty-one interconnected buildings spread over forty-plus acres in rural Tennessee, it housed an elabo-

---

* The endorsement came with a caveat: The body recommended giving higher priority to an enrichment method developed by the Berkeley scientists than the committee had proposed.

rate mechanical labyrinth involving hundreds of miles of pipe and tens of thousands of filters, seals, gaskets, and pumps—virtually all of which needed to be hex resistant.[35] To overcome the lingering technical barriers, Urey's team recruited more chemists. Even with the added government resources, DuPont couldn't manage to produce Teflon in the quantities needed.* But it brought other fluorocarbons from research bench to mass production at a previously unimaginable speed. By late 1943, DuPont had more than a thousand workers and several factories pumping out tens of thousands of pounds of these substances at its Chambers Works site in Deepwater, New Jersey.[36]

Fluorocarbons were hardly the only substances undergoing rapid development during this period. Faced with shortages of natural materials like steel and rubber, in 1941 the board overseeing U.S. military provisions had called for substituting plastics whenever possible.[37] The government had since spent huge sums developing synthetic materials and expanding the assembly lines of DuPont and other companies so they could produce the quantities needed for global warfare. As a result, onetime laboratory curiosities like synthetic rubber and polyethylene were suddenly being produced in massive quantities and fashioned into everything from bazooka barrels and parachutes to fighter-plane windshields. Charles Stine marveled at the fruits of this unprecedented collaboration between industry and government in a speech before the American Chemical Society: "The pressures of this war are compressing into the space of months developments that might have taken us a half-century to realize."[38]

But while most of the new synthetics grew out of established branches of chemistry, fluorocarbons were a virgin frontier mined

* Because Teflon warped under pressure, the modest quantities DuPont did produce weren't suited to uranium enrichment. But the material found other wartime uses, including as linings for liquid fuel tanks and nose cones for "proximity bombs."

with poorly understood hazards, and the frenzied pace left little time for developing safeguards. At DuPont's Chambers Works, the dangers of the fluorocarbon processing areas were legendary. Fires and explosions were commonplace; employees were regularly hospitalized with breathing problems, chemical burns, or worse. Manhattan Project inspectors warned their supervisors that widespread fear of injury was leading to unrest among workers and that DuPont employees had come to dread an assignment there as "an exile to Devil's Island."[39]

Workers weren't the only ones affected, though. In 1943, farmers downwind of Chambers Works began complaining that fumes from the plant were "burning up" their peach crops. There were reports of pastures littered with dead cattle and chickens, of farmhands vomiting all night after eating the produce they'd picked, of horses too crippled to work the fields. Some cows were so weak that they couldn't stand; they grazed by scooting on their bellies.[40]

The farmers' complaints alarmed Manhattan Project officials, who feared they would sue and that the government would be forced to cover damages under its contract with DuPont.[41] But Groves had a plan to contain this threat. That spring, he launched a secret division to study the health effects of "special" bomb-project materials, including radioactive fuels, and assigned a gifted young scientist named Harold Hodge to oversee the toxicology work.[42]

Based at the University of Rochester, the division occupied a six-story brick fortress that was accessible only by underground tunnels. As part of its mission to "strengthen the Government's interests" by producing data to defend against "medical legal" challenges, the division immediately organized a clandestine conference on the health effects of fluorine compounds.[43] Before the event, Conant, the Harvard president, reviewed the existing studies on file with the U.S. Public Health Service. He then sent an urgent letter to colleagues warning about the "extraordinary" toxicity of certain fluorocarbons.[44] There is little public record of what happened at the actual conference, held on January 6, 1944. The minutes are missing from the files at the U.S. National Archive, though an attendee list records the presence of Manhattan Project officials, a DuPont executive, and researchers from both the University of Rochester and Kehoe's lab at the University of Cincinnati.[45]

One week later, a "chemist's helper" at DuPont's Deepwater plant emptied a cylinder filled with a gaseous Teflon by-product into a ventilating hood. Some of the vapors wafted back into the room, where two other people were working. All three workers wound up in the hospital with breathing problems and cyanosis, a bluish tinge to the skin caused by lack of oxygen. The chemist's helper—who had been coughing up what DuPont documents described as a "thick, white, tenacious" mucus—eventually died, as did another victim.[46]

The Manhattan Project's medical division requested Teflon samples so it could investigate the cause of death, but DuPont refused to supply them. DuPont "considers that we were buying a 'packaged product' and is not interested in our investigating the toxicity of the materials involved," read an internal February 1944 memo to the division's director of medical research.[47] A Manhattan Project cable from the following month spelled out the company's rationale in starker terms: "DuPont is reluctant to release samples of their own commercially produced material since several of the components thus far identified give good promise for commercial uses."[48]

5

A Catch-22

The morning of Ersel Hickey's wake, Michael arrived at the funeral home carrying a large bouquet nestled inside Ersel's golf bag. He placed it near the casket, where his father lay dressed in jeans and his orange Cleveland Browns sweatshirt. Taking Sue and Katy's hands, Michael lingered there for a few minutes in shell-shocked silence. Then it was time to open the doors.

As he stood in the entrance shivering in his thin orange blazer, Michael saw the line of people waiting to pay their respects. It snaked down the block, past the faded clapboard rectory and the half-empty parking lots and Trustco Bank, before disappearing around the corner down Elm Street. Over the next five hours, a thousand or so people shuffled through the funeral home in their thick coats and muddy boots, their cheeks red from the long wait in the bitter cold. Some reminisced about Ersel giving them baseball caps when they were children or picking up their tabs when they ran into him at restaurants. Later, following a packed memorial service at the Immaculate Conception Church, Ersel's school bus and his yellow Corvette led a long procession of cars through town to the cemetery.[1]

After the funeral, blue oval magnets bearing Ersel's name started cropping up on cars and school buses all over town. (The Hickeys suspected one of the families on his bus route was behind this tribute, but they never found out for sure.) The county legislature passed a resolution memorializing Ersel, and while he didn't get to march in that year's St. Patrick's Day parade, he was honored with a giant banner atop one of the floats.[2]

Over the next few months, Michael found himself dreaming of his father every night and thinking about him constantly throughout the day. Every time he drove by Stewart's or the local golf course or the McCaffrey Street plant, he was flooded with memories. As the first in his family to finish college and work a white-collar job, Michael had sometimes been embarrassed by his working-class roots—when his father arrived on campus one semester in his "shit box" RV, Michael was mortified. And he'd never understood why Ersel cared so much about the kids on his bus or went out of his way to help people like the bartender at a run-down local watering hole.

But the outpouring of generosity toward his father moved Michael to the core. So many people had gone out of their way to support and honor Ersel, from the important folks in town to those who had almost nothing. People Michael wouldn't normally pay much attention to, if he was being honest. It struck him that his father had been right all along: It wasn't wealth or status that mattered; it was people. He resolved to be more like Ersel—more generous, more devoted to his community. On what would have been his father's seventy-first birthday, Michael walked into Stewart's, handed the cashier some money, and posted a sign dotted with Ersel photos and clip-art coffee cups. "A cup of coffee shared with a friend is time well spent," it read. "First 71 cups are on Ersel today!!!!! (Offer is contingent upon winking and/or smiling at one person . . .)"

At the same time, Michael was struggling to come to terms with Ersel's excruciating death. He was plagued by nightmares of his father thrashing in his hospital bed. And he was haunted by the feeling that his family had been cheated. Michael had been cheated out of a cherished bond that had taken years to form. Oliver had been cheated out of a relationship with his grandfather. Ersel had been cheated out of the retirement he had worked so hard for. Underneath all of this was guilt. Michael kept wondering whether Ersel would have lived longer or suffered less if he had steered him toward different treatments or fought to keep him on Votrient.

He also began to question why Ersel had even gotten sick in the first place. Practically everyone he knew in Hoosick Falls had stories about family or friends who had died young of cancer. In the case of factory workers, people sometimes chalked it up to hard living, but Ersel hadn't touched alcohol or cigarettes for more than thirty years. Oddly, Ersel's mother had been diagnosed with kidney cancer in her sixties, too—and it wasn't a type that was supposed to be passed down genetically. Ersel's brother, who had worked alongside him at the plant, also had potentially cancerous spots on his kidneys. Was it possible, Michael wondered, that all of this was something more than bad luck?

His mind raced with connections: the barrels in his parents' attic, the burning-rubber smell that wafted through their windows on summer evenings, the waxy film that coated the brambles where he used to pick berries. He thought about the rumors of clandestine dumping and the confirmed chemical spills he'd heard about over the years. In the early aughts, one of the former Dodge plants had dumped huge quantities of acid into the river, turning it an eerie chemical blue and killing hundreds of thousands of fish.[3] Michael looked into the chemicals used at that site, but none seemed to fit the pattern of disease he was seeing.

Then he got to thinking about the McCaffrey Street factory,

which was now run by a giant French corporation called Saint-Gobain. As he recalled the putrid sludge from the sinks where workers rinsed their Teflon dip pans and the crust he scrubbed out of the smokestacks, a horrid thought struck him. Maybe something dangerous was coming out of the stacks and getting into the wells in the adjoining field—the very wells that supplied the entire village's drinking water.

AROUND THE FIRST ANNIVERSARY of Ersel's death, one of Michael's oldest friends, a beloved high school math teacher, died of cancer at the age of forty-eight. After her funeral, Michael holed up in his home office to call a mutual friend, and they wound up talking about all the people they knew who had died young of cancer. Michael, whose research habits had grown increasingly compulsive, began punching keywords into Google. When he typed "Teflon" and "cancer," he came across a detailed study involving sixty-nine thousand West Virginia residents who had drunk water polluted with a chemical called perfluorooctanoic acid, or PFOA, which was used to make Teflon. The authors had found a "probable link" between the substance and six serious health problems—among them kidney cancer.[4] Michael felt his stomach drop.

He called out for Angela, hoping she would help him sort through the rush of thoughts and feelings. Instead, she gave him a despairing look. It wasn't so much the specter of pollution that worried her as the idea of challenging Saint-Gobain. Her family had lived in the area for generations, working at the plants and farming an old Civil War battlefield. Not only did she know everyone, her parents also knew everyone's parents and grandparents—a densely woven, multigenerational fabric that she held dear and that was knit together partly by the factories. If Michael pointed a finger at Saint-Gobain, which these days was the mainstay for local workers, Angela knew the backlash would be ferocious. He would get dragged through the mud, and some of the muck would rub off on his family. Even Oliver was bound to take flak from other kids. For their son's sake if nothing else, Angela begged Michael to let the matter go.

When Michael told other family members, their reactions were similar. His brother, who worked at a local gasket factory, was especially worried about Sue. An elderly woman living right by the McCaffrey Street plant would make an easy target for angry workers.

Michael agonized over what to do next. If there was any chance that other people were at risk of suffering like his father, he had to do *something*. At the same time, he was scared of making a mistake that would hurt his family or his community. So he delved deeper into research. Often, he sat up until four in the morning, poring over scientific studies, some of them so technical he had to read them four or five times before they made sense.

After several months of this, Michael was confident enough in his theory to approach the family doctor, Marcus Martinez. Since Ersel's death, Marcus had become one of Michael's closest confidants. It was Marcus he called late at night when the grief felt unbearable. And there was no one Michael trusted more when it came to protecting the community's health.

Marcus's family had been running the village's main medical practice since the 1950s. The elder Dr. Martinez had been known for his gruff bedside manner—when diabetic patients ate too much sugar, he'd put them on the phone with the undertaker to make their "arrangements." But what he lacked in delicacy, Old Doc made up for in devotion. Patients were treated on a first-come-first-served basis, and he stuck around the office until everyone was seen, even if that meant working until 10 p.m. Those who lacked money for his fees paid with whatever they could muster, be it carpentry or fresh eggs.

By the time Marcus took over the practice in the 1990s, the rise of for-profit HMOs had made assembly-line medicine the norm. But Marcus, a slight man who stood just over five feet tall, worked hard to deliver the same kind of personalized care his father had. In his office, sick patients were seen the same day, and every call was answered by a human being. "My philosophy was—and it was really pretty simple—bring them in if there's any question," he said. "When my head hits the pillow at night, I don't want to worry, 'Oh Jesus, we probably should have seen them.' "

Outside office hours, Marcus made house calls, sat vigil at people's deathbeds, and did rounds at the local hospital, allowing him to oversee almost every aspect of people's care. His patients trusted him absolutely and took comfort in his calm, direct manner.

As it turned out, Marcus already suspected that Hoosick Falls had a pollution problem because an unusually large share of his patients seemed to develop cancer at a young age or get rare, aggressive forms. After reading up on PFOA, he thought there might be something to Michael's theory. So the two men decided to approach the mayor, David Borge, about testing the water.

Borge, a soft-spoken retiree with wavy gray hair, had spent his career administering state programs for people with disabilities.[5] But he hadn't taken much interest in local government until a few years earlier, when a conflict with a neighbor over garbage inspired him to start attending village board meetings. He found them surprisingly uplifting. "I didn't see Democrats or Republicans," Borge recalled. "I saw a group of people trying to do the right thing and solve the community's problems together." Some board members, including Michael Hickey, who was serving alongside his father at the time, were impressed with his thoughtful questions. In 2011, when Michael gave up his seat, he asked Borge to replace him. One year later, Borge ran unopposed for mayor, a part-time job that paid about $11,000 and involved mostly mundane matters like snow plowing and sewer bills. Only fifty-two of the village's roughly three thousand residents voted in the election.[6]

Still, Borge took the job seriously. One of his top priorities was reversing the village's economic decline. Since the 1990s, when manufacturers began shifting production overseas or south of the Mason-Dixon line, most of the factory jobs had trickled away and the lion's share of local businesses had shut down, their stately Victorian storefronts vacant and crumbling.[7] Saint-Gobain had stayed. Its three local plants were just a tiny fraction of its global operations, with headquarters near Paris and 160,000-plus employees spread across seventy-six countries.[8] But for Hoosick Falls, where the firm was the largest private employer and the leading source of tax revenue, these factories were a lifeline.

Borge had been negotiating with the company to expand its local operations. At the same time, he was working with a new nonprofit on a raft of proposals to attract investment, among them converting vacant factories into low-cost space for startups and turning unused fields into marijuana farms. "No idea was too crazy," Borge recalled. Already, the team had begun work on a mixed-use development on a vacant lot downtown.[9]

Meanwhile, outside entrepreneurs were buying up abandoned buildings and turning them into vibrant new businesses. A ramshackle old paper mill had been transformed into a stylish brewery-restaurant, with spare modern decor celebrating the town's industrial history. A pair of nineteenth-century buildings in the village center were being reborn as a gallery and loftlike apartments. The once-decrepit armory was morphing into an arts and culture nonprofit offering after-school programs and meditation circles. A column in the region's largest paper marveled at the improbable renaissance that was turning Hoosick Falls into the "next up-and-coming downtown."[10]

It was against this backdrop that Michael and Marcus arrived at town hall with a stack of scientific papers and a story about some chemical with a convoluted name in the water supply. The mayor found the details baffling—but he immediately understood that a contamination scare could torpedo the village's economic plans.

Wary of alarming residents, Borge convened a closed-door session with the village board under the guise of "personnel and security issues," which are exempt from the state's open-meeting laws. He also sought guidance from the state and county health departments. He didn't get much. In their emails, officials from the two agencies—some of whom were clearly flummoxed by Borge's inquiry—briefly debated how to go about testing for the chemical. In the end, they concluded that it wasn't necessary.[11]

When Michael heard that the mayor had no plans to test the water, he decided he would do it himself. Using $500 of his own money, Michael bought a kit from AXYS Analytical Services, a Canadian company that had analyzed water for the West Virginia study.[12] Just before Labor Day 2015, a package containing a blue Igloo cooler, some ice packs, and four collection vials arrived on his doorstep. The directions called for collecting two raw and two treated drinking water samples, so Michael arranged a meeting at his home with the water plant operator, Jim Hurlburt, whom he'd worked with during his time on the village board. Michael asked Hurlburt to collect samples from the plant. When Hurlburt suggested clearing it with Borge, Michael called the mayor, who promptly shut him down. Private citizens had no business with the raw drinking water, Borge said.[13]

By then, Michael was hell-bent on seeing the process through, so he found a way to collect the samples himself, too. For the treated ones, he took water from his and Sue's homes, which were on the public system. As a stand-in for raw water, he decided to collect from local businesses that drew water from untreated private wells. First, he snuck through the side door of the local McDonald's and filled a vial from the bathroom faucet. Then he dispatched a reluctant Angela to the Dollar General across the street with a vial tucked in her purse to collect water from the ladies' room. Finally, he put the samples and ice packs into the cooler and shipped them back to AXYS.

. . .

JUST A FEW DAYS LATER, Michael was chatting on the phone with Marcus about an upcoming golf game when the doctor suddenly gasped and went silent. Intense stabbing pains were coursing through his shoulders and gut. Marcus got off the phone, climbed onto his exam table, and pressed his fingertips into his abdomen, looking for a rupture. He didn't find anything abnormal, so he decided to go to Southwestern Vermont Medical Center, the hospital where he did his rounds, for an ultrasound and a CT scan. He was lying alone in a hospital room later that evening when a colleague came to inform him that the scans had turned up two masses: one on his liver, one on his lung. A few days later, a biopsy confirmed Marcus's worst fear: cancer.

Marcus's mother, Gloria, a registered nurse who had worked in the family practice for decades, wasn't usually one for tears. But when Marcus told her the news, she broke down sobbing. "She didn't have the luxury of deluding herself," said Marcus's brother, Jamie. "In our family, when we heard someone had cancer in the liver and lung, we knew it was a death sentence."

The doctors in Vermont had diagnosed Marcus with small-cell lung cancer, meaning the prognosis was especially dire—most patients lived only a few months after the disease reached the liver. Marcus had his doubts. After all, he was only forty-two and didn't have any of the crippling symptoms he usually saw in lung cancer patients. He suspected he actually had a rare cancer called carcinoid, which attacks the neuroendocrine system. Unlike small-cell cancer, it could sometimes be rooted out through surgery. Marcus arranged to get a second opinion from a doctor at Dana-Farber Cancer Institute in Boston, one of the nation's leading cancer hospitals. However, when he arrived for his appointment, the oncologist hadn't even seen his file. The man stepped out for a few minutes to consult the records. When he returned, he confirmed the diagnosis and told Marcus that he only had about six months left to live.

But Marcus still had one more specialist to consult: his oncologist friend Eric Pillemer. As a medical student, Marcus had done a rotation in Pillemer's department at an area hospital and had been impressed by his dedication. In his own practice, Marcus worked closely with Pillemer when treating cancer patients and often sought his advice on medical quandaries. "I could run anything by him," Marcus said. "It didn't have to be oncology. He must have read every goddamn textbook in medical school, because he always could come up with something."

After discussing the details with Marcus, Pillemer agreed with his old friend's assessment: The lung cancer diagnosis didn't fit. Within a day, the oncologist had called around to several cancer centers and come up with a treatment plan: four rounds of chemotherapy to shrink the tumors, followed by surgery to remove them. Marcus knew a painful fight lay ahead, but at least he now had some hope that he would live to see another birthday or two.

THE VERDICT ON THE VILLAGE WATER arrived with little fanfare—a plain white envelope tucked inside a wad of advertising circulars. Michael fished it out of his mailbox one crisp October evening and began deciphering the data. Sure enough, all the water samples—including his and Sue's tap water—were heavily polluted with PFOA.*

When Michael told Sue the news, all she could think about was the brown Tupperware pitcher of tap-water-brewed iced tea that Ersel used to keep in the refrigerator. He drank a half dozen or so glasses a day, because he believed staying hydrated would keep him from getting kidney cancer like his mother had. It crushed Sue to

---

* Michael's tap water contained 540 parts per trillion. Sue's had 460 parts per trillion, well above the Environmental Protection Agency's then safety guideline of 400 parts per trillion, which was not enforceable.

think that his efforts to stay healthy had only added to the glut of toxic chemicals in his body.[14]

Michael, on the other hand, was somewhat relieved. Now he had proof that there was something wrong—surely the mayor would have to take him seriously. He and Marcus arranged another meeting with Borge. Michael explained their findings and handed the mayor a manila folder with the sampling results and data from the West Virginia study. To their surprise, Borge revealed that he had quietly commissioned his own analysis, which had also detected PFOA.[15] The mayor slid a copy of the results across the table—then launched into another spiel about the village's economic plans. Marcus could hardly believe what he was hearing.

Digging into Borge's data that evening, Marcus discovered that the numbers were just as concerning as the ones from Michael's improvised tests. One of the village wells had PFOA levels that were far higher than the safety guidelines set by the Environmental Protection Agency. And all of them were in the range associated with cancer and other diseases in West Virginia. Horrified, the doctor called Borge and begged him to take the village off the municipal water. "Dave, I'm telling you, you do not want your grandchild drinking this water," he pleaded. "It's sick, ethically sick, to let people consume this."

The mayor was still reluctant to do anything so drastic, but behind the scenes, he was searching doggedly for solutions. He directed the town's water plant operator to shut down the most polluted of the three village wells and conferred with the engineering firm that had designed the plant about methods for filtering out PFOA. He also had his deputy, Ric DiDonato, a jazz singer who performed with local big bands, look into the research on health effects.[16] Borge's pleas to state officials, meanwhile, took on a fevered urgency. "As you might expect, the Village Board of Hoosick Falls is very anxious to receive direction regarding the discovery of PFOA in our municipal water supply," he wrote in one email. "I am asking you to expedite any and all direction, suggestions and information which can assist our community in resolving this most troubling situation. . . . We need your help."[17]

Unfortunately, the authorities on the receiving end were working under the flawed assumption that the lack of federal regulation meant that there was no clear evidence that PFOA was harmful. As a result, they saw the pollution as primarily a PR problem. Rather than dealing with the contamination itself, their advice to Borge focused on managing the potentially "messy" public reaction.[18]

Eventually, Borge decided to appeal directly to Saint-Gobain. And in late 2014, he met privately with company officials, who argued it was highly unlikely that their firm was even responsible for the pollution, since it had never emitted PFOA from the McCaffrey Street stacks.[19] (In fact, Saint-Gobain and a subsidiary company had done so for years.)[20] Two weeks later, Borge and DiDonato visited the McCaffrey Street plant. There, they sat down with a plant manager and the director of research for Saint-Gobain's entire multimillion-dollar U.S. division. The executives spoke about the firm's contributions to the village and its plans to expand its local operations.*[21]

These gatherings set the tone for the negotiations that followed. In-person meetings were often held on Saint-Gobain's turf, with the retiree, and sometimes his jazz-musician deputy, trying to hold their own against a team of high-level executives. Both sides agreed not to involve lawyers and to exclude most other village officials. That way,

---

* Saint-Gobain declined to answer specific questions for this book, but it issued a statement saying, "Saint-Gobain is, and has always been, committed to acting as a responsible environmental steward in the communities in which we live and operate, which includes complying with regulatory requirements. . . . Any PFOA that has been used in our production originates from raw materials purchased from our suppliers, whom we rely on to follow industry laws and regulations, with the suppliers' conformity in this regard reviewed as a component of our product stewardship process."

the board didn't have to inform the public under the state's open-meetings laws.[22]

This approach succeeded in keeping negotiations friendly—in their emails, company executives thanked Borge for "fostering a spirit of cooperation."[23] But it also made it easy for Saint-Gobain to stall meaningful action. As a gesture of good faith, the company agreed to investigate the source of the pollution. Until the results were in, however, it refused to provide residents with bottled water, much less take steps to clean up the public water supply.

The mayor eventually tried to secure bottled water for distribution through the county's emergency program but was told the village didn't qualify, because the PFOA levels didn't violate any regulations.[24] "It was a Catch-22," Borge noted. "Because this was an unregulated contaminant all the regular avenues for help were closed."

IN THE FALL OF 2014, as Marcus was struggling through chemotherapy, he bought a burial plot and asked his brother to write his obituary in case the treatment didn't work. Although he was bone-tired, he also redoubled his push to get the village clean drinking water. After a life spent serving his community, Marcus didn't feel he could die in peace until he knew the water was safe.

During appointments, the doctor started warning patients outright not to drink the water. He also urged the village priest to shut down the drinking fountains at the local Catholic school, St. Mary's Academy, which was on the municipal system. (The public school, on the other hand, drew water from an uncontaminated private well.) When the priest raised the issue at a public village board meeting in late 2014, the board assured him this step wasn't necessary.[25] So Michael found his own way to get the word out. Over a round of golf, he explained the problem to an outspoken friend whose children happened to attend St. Mary's. Word quickly spread to other parents, some of whom sent Borge anxious emails or showed up in classrooms with cases of bottled water.

At this point, Borge drafted a letter to be distributed to residents with their water bills. It explained that a "synthetic element" had been detected in the pretreated municipal water, but suggested that it was being removed by the village's "state-of-the-art" filters.[26] (It wasn't.) The letter also claimed that the treated tap water met or exceeded "all County, State and Federal standards for public health safety," even though Borge knew otherwise; the village had recently commissioned testing, which found that the PFOA levels blew past the EPA's safety guidelines.[27]

When a draft of the letter was posted to the village website, Michael and Marcus were outraged. Hoping to keep it out of mailboxes, they turned up at the December village board meeting in the drab conference room that doubled as the village courthouse. Many of the folks there were Marcus's patients—people he'd seen through minor ailments and major health crises or comforted during the death of a loved one. With a knitted cap pulled over his patchy, chemo-ravaged scalp, the doctor stood up and implored the trustees not to send out the letter and, more pointedly, not to wait on outside agencies to intervene. The water was toxic; people needed to stop drinking it immediately.[28] The trustees listened intently, but they were not swayed. "I respect Marcus a lot—I think all of us do," one longtime board member later said. "But the authority for us is New York State and Rensselaer County, and they were telling us the water was safe."

When the letter was sent out, some of the St. Mary's parents were outraged. One created an anonymous Facebook page and, with Michael's help, posted scientific studies linking PFOA to a host of serious diseases. Another parent called the regional EPA offices. Although two months had passed since Michael had shared his sampling results with Borge, this was the first the agency had heard of the matter. In response, a regional EPA supervisor emailed a contact at the state Health Department, who confirmed that one of the village wells had PFOA levels above EPA guidelines but noted that this well had "reportedly been taken off line." With that vague assurance, the EPA dropped the matter.[29]

6

Biological Dynamite

The mysterious blight that had ravaged the countryside around DuPont's Chambers Works plant in 1943 had left the area's farmers utterly devastated. Most hung on to their land, hoping whatever was ruining their crops had run its course. But the following harvest season was just as disastrous. Across southeastern New Jersey, crops shriveled, animals foundered, and farmhands fell ill. Desperate, a group of peach growers hired a lawyer named William Gotshalk to investigate.[1] Both he and the farmers suspected the problem was fumes from the East Area of Chambers Works, where—unknown to them—DuPont was manufacturing fluorocarbons. So Gotshalk, a former prosecutor from Camden, New Jersey, requested information on the chemicals used there. When DuPont stonewalled, saying they belonged to a secret military program, the lawyer turned to a renowned analytical chemist named Philip Sadtler for help.[2]

After touring the orchards, Sadtler concluded that "practically all the vegetation" had been damaged. But the problems went far beyond that. At a grade school near the plant, he noticed that something had eaten away the surface of the windows. A cow in the neighboring field was so crippled it couldn't stand. Other pastures in the area were littered with dead chickens and horses. Sadtler also met farmers who were struggling with strange symptoms, including bouts of vomiting and muscle weakness.

The chemist began analyzing fruits and foliage from more than two dozen orchards at varying distances from the plant site, along with the blood of animals and farm workers. Many of the samples turned out to contain astronomical levels of fluorine.[3] Available testing methods couldn't pinpoint the specific compounds, but they seemed to support the farmers' suspicion about the source—the closer the produce was grown to DuPont's plant, the higher the fluorine levels.[4]

As he was wrapping up the investigation in the spring of 1945, Sadtler was summoned to his Philadelphia office. There, he was greeted by two men who turned out to be army counterespionage agents, and they accused him of "dealing with the enemy."[5] Sadtler didn't know it, but by then the Manhattan Project was in its frenzied final phase, with more than 120,000 workers around the country racing to finish the bomb. America's military leaders believed that a surprise nuclear attack on Japan was the key to winning the war, so maintaining secrecy was especially crucial.[6]

They proved their theory with breathtaking ferocity on August 6, 1945, when a U.S. plane dropped a uranium bomb on Hiroshima, killing about 140,000 people. Three days later, a plutonium bomb hit Nagasaki, killing another 70,000 and effectively ending the war.[7] The New Jersey peach growers had avoided litigating during the conflict out of patriotism, but now they immediately brought suit against DuPont—the first legal action stemming from the bomb project. Once again, Gotshalk demanded information on the chemical used in the East Area. This time, Manhattan Project officials blocked the release, citing the "nation's military security."[8]

The lawyer never managed to pry loose the data, but Sadtler did inform the Food and Drug Administration about the tainted produce. After performing its own analysis, the agency moved to bar the sale of fruits and vegetables from the region. Fearing this would cause a PR crisis, not to mention more lawsuits, DuPont pressed Manhattan Project brass to intervene.[9] In early 1946, two members of General Groves's staff met with the chief of the FDA's food division, alerting him to "the substantial interest which the Government had in [any legal] claims which might arise" from the FDA's actions. After that, the agency shelved its planned embargo.[10]

Groves himself saw the peach crop litigation as a grave threat to America's expanding nuclear weapons program and was thus eager to mollify the growers. Just after the FDA meeting, he invited Willard Kille, the chair of the farm labor committee for one of the affected counties, to dine with him in Washington.[11] The general promised to investigate the blight and compensate farmers for any portion caused by the bomb project. Kille responded with an effusive letter of thanks. "I wish all of our farmers could have been present," he wrote. "I think . . . they too would have come away with the feeling that their interests in this particular matter were being safeguarded by men of the highest type whose integrity they could not question."[12] Behind the scenes, meanwhile, Groves was quietly marshaling the vast resources of the federal government to crush the growers' legal case.

One of Groves's deputies was dispatched to Chambers Works, along with Harold Hodge, the chief toxicologist for the bomb project's secret medical division.[13] By then, the division was operating under the auspices of the Atomic Energy Commission and had hundreds of employees, many of them engaged in ethically dubious research. (One of Hodge's chief projects involved injecting unknowing cancer patients with plutonium and would later be the subject of a major federal investigation.)[14]

During the New Jersey trip, Hodge, the deputy, and several DuPont officials toured the plaintiffs' orchards and debated strategies for derailing the farmers' claims. Drawing on the delegation's ideas, in the spring of 1946 Groves launched an elaborate research program.[15] According to declassified government records, various agencies were enlisted to help deflect blame from Chambers Works, with the goal of protecting "the interests of the Government at the trial of the suits brought by owners of peach orchards." The Department of Agriculture and the Chemical Warfare Service were charged with analyzing the region's air and soil in the hopes of sowing "doubt as to the origin" of the pollution and possibly pinning blame on other factories.[16] Hodge and the Public Health Service began an ambitious campaign to persuade communities across the entire country to put fluoride in drinking water, touting the benefits for dental health. Their real goal, though, was to convince the public that fluorine compounds were safe, since they played a crucial role in America's expanding nuclear weapons program.[17] Hodge's team also started to analyze the substances Sadtler had detected in the blood of local residents, trying to figure out exactly which compounds they were dealing with.

Both Sadtler and government officials suspected the main culprit was hydrogen fluoride, a highly toxic gas used in various bomb-project materials. Hodge's lab began studying how rats metabolized this substance.[18] DuPont cosponsored similar research on human subjects at Robert Kehoe's lab. During one experiment, an African American lab assistant and a staff toxicologist were placed in a clear plastic chamber with varying quantities of hydrogen fluoride for six-hour daily blocks, stretched over a period of six or more weeks. While the researchers recorded obvious symptoms like skin irritation, they made no effort to assess the effect on the two men's health; they simply tracked how the chemical moved through their bodies. It turned out they expelled it rapidly in their urine, which prompted the question: What *was* the mystery substance accumulating in the blood of so many people and animals around the DuPont site?[19]

By 1947, Hodge was one step closer to an answer. There were, he discovered, two very different fluorinated substances detectable in people's blood. One was the fluoride ion, which occurs naturally in certain rocks and clay and, thanks largely to Hodge, was being put in a growing share of the nation's drinking water. The other was fluorocarbons, which were almost exclusively man-made. Hodge directed his team to refocus their research on these substances, which he'd concluded were "more toxic." The results, he noted, would be relevant not only to the farmers' lawsuit but also to ordinary people who could be exposed through "heavily contaminated" food.[20]

There is little public record of the team's findings. In the late 1940s, the Atomic Energy Commission declassified virtually all Manhattan Project papers with no direct bearing on national security. But it held back those that dealt with medical research or pollution near project sites, arguing they could tarnish "the public prestige" of the U.S. government and "encourage claims against the Atomic Energy Commission and its contractors."[21]

WHILE HODGE WAS BUSY INVESTIGATING just how toxic fluorocarbons were, private companies were racing to transform them into products they could market to the public. One of the most active players was Minnesota Mining and Manufacturing Company, or 3M. Around the time the secrecy order on fluorocarbons was lifted in 1945, the Saint Paul–based firm acquired patents for the technology to produce them and hired a team of former Manhattan Project chemists to turn them into new compounds.[22] It was a bold move for a company whose biggest innovation to date was waterproof sandpaper, but its leadership could see the technology's revolutionary potential. In the past, virtually all man-made materials (and many natural ones) had been composed primarily of two elements: carbon and hydrogen. This simple combination had spawned thousands of compounds, which had been fashioned into a plethora of consumer goods, from appliances to denture cream. Adding fluorine to the mix could lead to a whole new universe of even more transformative substances, thanks to the properties conferred by the tenacious carbon-fluorine bond. 3M scientists envisioned wonders like cars with sealed-in oil that never needed changing and fireproof paint that protected houses from burning down.[23]

Over the next few years, the team turned out countless novel compounds, most of them belonging to the same previously scarce class of chemicals that had been the focus of Manhattan Project chemists: PFAS. "Almost every day, we made a new molecule which had never been on the face of the earth before," one 3M scientist marveled.[24]

But despite their astonishing properties, nobody involved knew how to turn these innovations into actual products for consumers.[25] So in the late 1940s, 3M began publicizing its creations in the hopes that other companies would find ways to use them. ("WANTED—Jobs for a Trillion New Chemicals," read a headline in *Popular Mechanics*.)[26] A supervisor from the fluorochemical division pitched them to manufacturers around the country. Each time, he would close by saying, "Now you know something about these unusual compounds—all the possibilities they offer for solving problems that have been intractable up to now. So do you have any problems?"

As it turned out, the chemical giant DuPont had a big problem: It was still struggling to overcome the hurdles that had made Teflon so difficult to work with. But insiders suspected the issue could be solved by a processing aid with the right qualities. So in 1950, they approached 3M, which suggested PFOA. Sure enough, when Teflon particles were mixed with a PFOA solution, they yielded a substance that could be more easily molded or applied as a coating.[27] The same year, DuPont began mass-producing Teflon at its sprawling new plastics plant near Parkersburg, West Virginia.[28] 3M, meanwhile, introduced its revolutionary fabric protector, Scotchgard, which contained ingredients that broke down into a close cousin of

PFOA, known as PFOS.[29] The product gave ordinary textiles uncanny resistance to grease and stains and soon found its way into thousands of household items.

THE PUSH BY COMPANIES LIKE 3M to turn wartime innovations into peacetime profits would transform American life. After the conflict, manufacturers began marketing these materials for every imaginable purpose. Poison gases found new life as pesticides. Explosives like ammonium nitrate were repackaged as chemical fertilizer, revolutionizing entire food systems.[30] And plastics once reserved for military use were transformed into a cornucopia of goods. Polyethylene, which had been used to coat radar cables, was turned into Tupperware, Hula-Hoops, and grocery bags. Vinyl, or PVC, became shower curtains, flooring, medical equipment, and a popular new household item called Saran wrap.[31] Nylon, which had been used to make parachutes, flak jackets, and aircraft fuel tanks, returned to store shelves in the form of previously scarce runproof stockings. Across the country, "nylon riots" erupted as thousands of women blocked traffic, smashed shop windows, or stampeded into stores, toppling shelves and sometimes other customers. ("Women Risk Life and Limb in Bitter Battle for Nylons," reported one Georgia paper.)[32]

With the world suddenly awash in synthetics, the horizons of human possibility were no longer limited by a finite supply of natural materials. Man-made substances were so versatile and could be produced in such large quantities that people now had access to a huge variety of low-cost goods. And this abundance reshaped American society, ushering in an era of disposability and rapid technological progress. It also fueled the postwar economic boom. As author Susan Freinkel observed in her book *Plastic: A Toxic Love Story*, America became a nation of consumers "increasingly democratized by our shared ability to enjoy the conveniences and comforts of modern life. Not just a chicken in every pot, but a TV and stereo in every living room, a car in every driveway."[33]

The synthetics revolution also brought thousands of new chemicals into American homes. Most people didn't give much thought to the implications, but manufacturers flocked to Kehoe's lab to sponsor research on the substances they were using. His files brimmed with unpublished reports on the hazards of materials like vinyl and DDT, which would become omnipresent in the postwar years.[34] Kehoe believed the secrecy was justified. Synthetic materials, he argued in a 1963 essay, would be desperately needed to "feed, clothe and house those who will populate [this] land in succeeding generations." Given that the science was still developing, it was "neither wise nor kind" to focus the public's attention on their toxicity.[35]

But Kehoe didn't have a monopoly on this information, as he once had with the research on leaded gasoline. By the 1940s, other scientists had begun investigating the health effects of man-made chemicals. One of the most influential was a German-born pathologist named Wilhelm Hueper, who had served in the military regiment that unleashed the world's first mass poison gas attack, killing more than a thousand Allied soldiers during World War I.[36] After emigrating to the United States in 1923, he conducted research on environmental triggers for disease at various institutions before being recruited by DuPont's new industrial toxicology lab in the 1930s. There, he documented links between a number of commonly used chemicals and cancer, and sparred viciously with his supervisors over whether to make the findings public.[37]

Hueper went on to publish a landmark book making the case that man-made chemicals "were the most important and frequent causes of human cancers."[38] This work, which drew on a wide range of evidence, would help give rise to the fields of occupational and environmental medicine. By 1948, it had also propelled him into a job as founding director of the National Cancer Institute's Environmental

Cancer Section, where he continued his research.[39] In one study, he reported that inserting slivers of various plastics—including Teflon—under the skin of rats caused them to sprout tumors, which he attributed to the "slow release of very small amounts of chemicals."[40] Hueper's work did not please his former employer. DuPont sent a letter to the FBI accusing the German émigré of belonging to the Nazi Party, and another to the Cancer Institute claiming he had "communistic tendencies." In the age of McCarthyism, these were ruinous accusations. The federal loyalty commission began investigating Hueper for treason; his superiors at the Cancer Institute canceled his planned promotion and began censoring his papers and speeches.[41]

By the 1950s, however, other scientists were confirming Hueper's findings in published, peer-reviewed papers.[42] The emerging consensus was that many man-made chemicals had unusual biological potency, making them harmful at lower doses than ordinary poisons. At the same time, chemists wielding powerful new detection tools were discovering that these substances were widespread in the environment. And a small but vocal group of activists began raising concerns about the lack of testing for chemicals in the food supply.[43] They found an advocate in James Delaney, a Democratic congressman from New York, who publicly decried the food-safety law passed in 1938 as "a tragic legal joker that permits us to become a nation of 150,000,000 guinea pigs guilelessly testing out chemicals that should have been tested adequately before they reached our kitchen shelves."[44]

In the early 1950s, Delaney formed a committee to investigate the issue of chemicals in food, including pesticides, additives, and residue from packaging. He called Hueper as a lead witness. During his testimony, the pathologist stressed that many man-made chemicals could wreak havoc on the human body—especially carcinogens and synthetic hormones, one of which he likened to "biological dynamite." Some of his colleagues had found that only "four-tenths of a thousandth of a milligram" of a certain chemical could increase the incidence of cancer in mice. And because the effects were cumulative, Hueper argued, no level of exposure could be presumed safe. He advised the lawmakers to require that chemicals in food be "tested for toxic and possibly carcinogenic properties," and to ban those that cause cancer. Other prominent scientists echoed his plea.[45]

But the titans of American industry had other ideas. Aided by Hill & Knowlton, the PR firm that would later pioneer Big Tobacco's campaign to discredit the science on the harms of smoking, chemical interests launched a major offensive to fend off testing requirements. They lobbied lawmakers, hosted all-expense-paid conferences for journalists, and put pro-industry science materials in thousands of public school classrooms.[46] The industry also planted news stories asserting that chemicals would lead to tastier food and a safer, more bountiful food supply, all but eliminating global hunger.[47]

These efforts paid off. In 1958, Congress passed a law requiring safety testing for chemicals that wound up in food, including residue from packaging and cookware. Those that were linked to cancer would be banned outright. But at industry's urging, the thousands of substances already in use, including Teflon and PFOA, were presumed to be safe and grandfathered in.[48]

IT WAS AROUND THIS TIME that an enterprising French engineer named Marc Grégoire introduced the world to the wonders of nonstick cookware. Grégoire had hit on the idea after coating his wife's muffin tins with Teflon to keep dough from sticking to the sides. It worked so well that in the mid-1950s, he formed a company called Tefal to sell nonstick pans. At first it was a modest operation, with Grégoire coating pans with Teflon in his kitchen and his wife peddling them on the street. But by 1958, Tefal had built its first factory and was selling 500,000 pans a year.

That summer, an American businessman named Thomas G. Hardie attended a party in Paris and happened to meet Grégoire, who regaled him with stories about his popular cookware. Hardie was convinced it would be a hit with American housewives. After returning from France, he met with DuPont executives to discuss the possibility of importing Tefal pans.[49]

Up until this point, DuPont had avoided marketing Teflon for use in home kitchen products because of toxicity concerns. Workers who inhaled Teflon fumes developed flu-like symptoms, and company doctors suspected exposure could cause far more serious problems.[50] Studies underway at Kehoe's lab lent credence to their fears. In 1954, scientists there had exposed dogs, guinea pigs, rabbits, and mice to the gases Teflon emitted when heated. Most of the animals died within minutes. Dissections revealed "diffuse degeneration of the brain, liver and kidneys," suggesting that something was getting into the animals' blood and attacking their organs. Kehoe concluded the culprit was probably other fluorocarbons that Teflon gave off as it broke down.[51]

Cookware rarely reached the kind of temperature used in these studies, but Kehoe suspected that similarly toxic chemicals could still leach out of Teflon pans during cooking. The year before Hardie's visit, when the Delaney hearings were still underway, a lawyer had approached DuPont about marketing Tefal pans in the United States. DuPont referred him to Kehoe. Anticipating strict new regulations, Kehoe discouraged him from proceeding. Many fluorocarbons had proven "exquisitely toxic," the scientist cautioned. Thus, "elaborate" and "very expensive" testing would be necessary to prove these chemicals could be safely used in cookware.[52]

But by the time Hardie made his pitch, the food additives bill had passed Congress. Because Teflon's ingredients had been grandfathered in, no elaborate testing was needed. So DuPont prepared to market Teflon products to the public. In 1959, it invited a reporter from *Popular Science* to its Wilmington, Delaware, headquarters for a pancake demonstration. According to the magazine, the cakes came out nicely brown and left no crusty residue, "because the pan was lined with Teflon, a remarkable fluorocarbon plastic" that was "as slippery as ice on ice."[53]

Tellingly, before the cookware had even landed on store shelves, DuPont was maneuvering to shield itself from legal liability. The company dispatched a chemist named Albert Henne, who had worked on the Manhattan Project's fluorocarbon program, to Kehoe's office to discuss a "rehabilitation campaign" for fluorinated substances in the "food business."[54] In a follow-up letter, Henne outlined a series of positions that DuPont wanted Kehoe to publicly stand behind: "I believe you share my belief that there is no practical risk involved. . . . I believe you may be willing to voice those opinions to a licensee if I formally introduce him to you. . . . I believe you might be willing . . . to act as an expert, either to supply him with data or act as a competent witness in the case of a lawsuit . . . if a suitable business arrangement can be arrived at." Kehoe agreed. "I shall be entirely willing to state my views to the people with whom you are consulting," he replied in October 1958. "I would be willing to support them (my views) in any formal manner in the courts or elsewhere."[55]

Kehoe made good on that promise. In early 1959, Hardie and a partner were granted rights to import Tefal kitchenware. On DuPont's advice, Hardie approached Kehoe for guidance on safety testing.[56] In stark contrast to his warnings just one year earlier, Kehoe replied that he could "see very little reason for concern about any public hazard." He discouraged Hardie from pursuing toxicology testing with the same ardor that he had recommended it in the past. "I am in considerable doubt that such investigations are necessary or will serve a useful purpose," he counseled. "Unless it is necessary to provide evidence to a government agency, such as the Food and Drug Administration, it may be both possible and wise to avoid such an attempt (which will not be very satisfying) and the expense associated with it."[57]

A DuPont salesman later accompanied Hardie on his first sales call, to Macy's Herald Square, and in December 1960, several hundred imported Tefal "Satisfy" skillets went on sale there. Despite a major blizzard, they sold out almost immediately. Soon Hardie was flooded with orders, and DuPont began marketing Teflon to American cookware makers. By 1961, the DuPont-branded Happy Pan was available at department stores across the country.[58]

While the first batch of Happy Pans were flying off store shelves, DuPont's chief toxicologist, Dorothy Hood, alerted executives to two troubling new in-house studies. Previous industry research hadn't pinpointed which of Teflon's ingredients or breakdown products were causing symptoms in workers and lab animals. But the new data raised a flag about the chemical PFOA—then known as C8 because of its eight-carbon backbone, swaddled in fluorine atoms. Animals exposed to the substance developed enlarged livers and kidneys—a sign that PFOA was toxic. Hood warned superiors it should be "handled with extreme care."[59]

DuPont never disclosed this information, but it ramped up internal safety testing, as did 3M. DuPont also took steps to keep PFOA from tainting drinking water around its plants, including dumping its Teflon waste at sea. But in the mid-1960s, after a fishing trawler dredged up a barrel, resulting in some embarrassing publicity, DuPont began secretly dumping Teflon sludge into unlined pits around its Parkersburg plant—and directly into the nearby Ohio River.[60]

Even as warning signs were piling up in corporate labs and offices, demand for fluorochemicals was skyrocketing. This was largely thanks to two players. One was the U.S. Navy, which during the 1960s collaborated with 3M to develop a highly effective firefighting foam that contained PFOA and precursors to PFOS. It would be deployed at tens of thousands of airports, fire stations, and military bases across the country.[61] The other was a former DuPont engineer named Bill Gore, who had worked on the crash program to develop Teflon for wartime uses and had a passion for tinkering and invention. In 1958, he founded a company from the basement of his Delaware home, using seed capital raised from his bridge club. At first, W. L. Gore & Associates specialized in Teflon-coated cabling for NASA missions and high-tech ventures like the IBM System/360, the first mass-produced digital computer. But Gore had even grander ambitions. If he could just figure out a way to unfold Teflon's scrunched-up molecules, he believed it would make for a tougher, more pliable substance with previously undreamed-of applications.[62]

Gore and his son, Robert, a fellow engineer, tried slowly stretching Teflon rods heated to various temperatures. But every time the rods reached a certain length, they snapped. Then, late one night in 1969, Robert grew frustrated and yanked the Teflon hard. This time it didn't break. *The New York Times* would liken the resulting material to "Edison's first lightbulb, the beginning of something altogether new and different." It had all the strength and flexibility that Gore envisioned. And because it had billions of tiny pores packed into every square inch (the result of air pockets emerging as the molecules unfolded), it was both watertight *and* breathable.[63]

During a ski trip to Vail in 1971, Bill Gore met the cardiovascular surgeon Ben Eiseman for cocktails, wearing a tie made from expanded Teflon. Eager to show off its properties, Gore smeared his tie with ketchup that vanished as if by magic when he wiped it with a napkin. He then covered his cupped palm with Teflon fabric and poured water into it. His hand stayed dry. Yet when he lit a match and blew at it through the same material, the flame flickered out, proving it was permeable by air.[64] Eiseman had been looking for a breathable material to repair damaged veins and arteries, so he asked Gore to send him samples for testing on pigs. And by 1975, Gore's company was producing a line of lifesaving vascular grafts.[65]

An avid outdoor enthusiast, Gore also took a handsewn Teflon tent on a family camping trip to Wind River, Wyoming. The first night, it rained and hailed, and the tent filled up with water. "Our sleeping bags were sopping," Bill's wife, Genevieve, recalled. The family eventually worked out the kinks in the material, and in 1976, W. L. Gore & Associates rolled out its trademark Gore-Tex—the world's first breathable, waterproof fabric.[66]

The material revolutionized Americans' relationship with the great outdoors. Unlike the heavy rubberized jackets of the era, which left wearers marinating in their own sweat, Gore-Tex parkas blocked out rain and snow *and* vented perspiration. This kept people warm for longer periods. Similarly, Gore-Tex tents and sleeping bags were lighter and more water-resistant than other camping gear, making outdoor adventures accessible to a far broader swath of the public.

In a sign of the material's amazing versatility, Gore-Tex was soon being fashioned into space suits, guitar strings, artificial ligaments, and computer cables that could carry signals at nearly the speed of light—far faster than any other material in existence.[67] It also found its way into everyday items like baby shoes and dental floss that pervade the most intimate corners of our lives.

7

Blood Secrets

Up until the 1960s, most Americans saw synthetic materials as an unqualified good—marvelous inventions that enabled rapid social progress. Then, in 1962, the naturalist Rachel Carson published *Silent Spring*. This lyrical and sharply reasoned book focused largely on the devastation that the widespread use of pesticides had wrought on nature. But it also introduced the public to the disquieting idea that man-made chemicals were inundating people's bodies. "For the first time in the history of the world, every human being is now subjected to contact with dangerous chemicals, from the moment of conception until death," Carson wrote.

What set these new chemicals apart from their naturally derived predecessors, she argued, was their "immense power not merely to poison but to enter into the most vital processes of the body and change them in sinister and often deadly ways." They destroyed the enzymes that protected the body from harm, disrupted the normal functioning of organs, and caused slow but irreversible change to our cells, a process that, over time, could lead to cancer.[1]

Most of the data Carson drew on wasn't new. It was the same

body of science that researchers like Wilhelm Hueper—whom Carson cited at length—had developed and synthesized decades earlier.[2] But Carson was the first to pull it together for the general public and lay out such bleak, far-reaching conclusions.

It is hard to overstate the influence of *Silent Spring*. It sold more than two million copies and prompted a tsunami of media coverage, along with two congressional hearings and a major investigation by the White House. The book also transformed Americans' attitudes toward the synthetic materials that now enveloped their daily lives. Suddenly, millions of people came to the stark realization that these supposed miracle substances were threatening the vitality of the natural world, a shift that gave rise to a broad-based environmental movement that transcended ordinary political divides. With nearly unanimous public support, the Nixon administration enacted a series of sweeping federal initiatives, culminating in 1970 with the formation of the EPA.[3] The following year, Congress drafted legislation giving the new agency the power to assess the safety of chemicals and restrict those that were found harmful. The Manufacturing Chemists' Association, as the industry lobby was then known, managed to block its passage.[4]

By the mid-1970s, however, a series of chemical-plant disasters was inspiring mass protests and lending new force to calls for government action. Congressman Robert C. Eckhardt, an eccentric Texas Democrat who bicycled around Washington with his files stuffed in a whiskey case, decided to seize the opportunity. He began pushing a stringent House version of the stalled toxic substances bill, which by early 1976 had amassed wide bipartisan support. The Manufacturing Chemists' Association responded by dispatching its chairman, a dapper DuPont executive named Richard Heckert, to shape the legislation in industry's favor. Every few days for the next several months, Heckert huddled with Eckhardt and his aides in the congressman's Capitol Hill office.

During tense early negotiations, the chief sticking points were whether industry should have to notify the EPA when it brought new chemicals to market and systematically test them for safety. Eckhardt wanted both provisions. But chemical concerns based in his district lobbied to kill them, and some lawmakers threatened to block the bill if its testing requirements were too tough.[5]

Hoping to avoid the kind of impasse that had doomed earlier legislation, Eckhardt compromised. Under the Toxic Substances Control Act, which passed Congress in October 1976, existing chemicals were again grandfathered in. Manufacturers had to inform the EPA when they introduced a new chemical and hand over any internal data finding that it posed a "significant risk." But they weren't required to proactively test it for safety.* This is why the vast majority of the 80,000-plus chemicals circulating in the United States today have *never* undergone any form of safety testing.[6]

OVER THE NEXT DECADE, Teflon became ubiquitous, gradually infiltrating everything from body lotion and carpeting to that most American of icons, the Statue of Liberty. (In the mid-1980s, a layer of Teflon was placed between its steel frame and copper skin to prevent corrosion.)[7] The material's inventor, Roy Plunkett, was celebrated as a visionary and invited to dine with Ronald Reagan, who had been branded the "Teflon president."[8]

Thanks largely to Teflon's success, DuPont's Parkersburg plant grew into a small city with its own fire department and medical center. In an area where few people had college degrees, the site was a rare island of opportunity. The thousands of men and women who worked there were known for their relative wealth and were often singled out for special treatment. Some people claimed that

---

* Technically, manufacturers were required to test if the EPA specifically requested it, but they could force the agency to defend the necessity of such testing in court simply by filing an objection, which effectively tied the agency's hands.

"DuPonters" could walk into a bank and get a loan just by saying they worked for the company.

For Sue Bailey, a ruddy, blue-eyed Parkersburg native whose father had worked at the company for years, the best part of being in DuPont's orbit wasn't the status or the lifestyle it afforded. It was the sense of belonging. With its family picnics and basketball tournaments, the plant felt more like a community than a workplace. So when the young mother got a job in the nylon division in 1978, she was overjoyed. Two years later, while newly pregnant with her third child, Bailey was assigned to the Teflon area. Working in a windowless concrete room, she channeled Teflon waste from a storage tank into an outdoor pit using a contraption that looked like a bicycle pump. Occasionally, the system overflowed with a thick green sludge that she had to slop up with a squeegee. During her previous pregnancies, Bailey hadn't experienced morning sickness, but this time she was constantly queasy and agitated.

When Bailey gave birth the following January, the baby had only half a nose and a serrated lower eyelid that gaped down to his cheekbone. Her doctor warned that her son, who was struggling to breathe through his misshapen airways, might not live until morning. Sue was so distraught that she couldn't bear to hold him. "I was terrified he was going to die in my arms," she said. "I asked the nurse to get my pastor, then I started screaming, and they had to give me a sedative."

Little Bucky survived the night. The following morning, he was transported to a children's hospital in Columbus, Ohio, where he would undergo the first of more than forty surgeries. Bailey, meanwhile, received a call from a DuPont staff physician asking about the baby's deformities. He claimed it was a routine inquiry. But after Bailey returned to work later that year, she happened to find a memo on the locker room bench. It described a recent 3M study that had documented "birth defects in the eyes of unborn rats" exposed to

PFOA in utero. Female workers who came into contact with the chemical were urged to consult their doctors "prior to contemplating pregnancy."[9]

Bailey took the paper to the on-site medical office and demanded to know whether the chemical had anything to do with her son's birth defects. The plant physician insisted there was no connection, but a few months later, a friend put her in touch with another Teflon worker named Karen Robinson, who had given birth to a baby with similar eye deformities.[10] "That pretty much clinched it for both of us," Bailey said.

Bailey continued working at the factory—she needed the insurance to pay for Bucky's numerous treatments. But she noticed the foremen were treating her differently. "I just wanted them to say, 'Yes, this is what happened to your baby and we're really sorry about it. Is there anything we can do to help you?' " she recalled. "Instead, they shunned me like I had the plague."

Her superiors had reason to be nervous. Seven years before Bucky's birth, two toxicologists who worked with Harold Hodge in the bomb project's medical research program had discovered a certain fluorocarbon accumulating in the blood of people around the country. Wondering what the mystery molecule was, in August 1975, one of the scientists, Warren Guy, placed a call to 3M and reached a chemist named G. H. Crawford. According to Crawford's memo about the call, Guy speculated that the seemingly " 'universal' presence of such compounds in human blood" might be related to the widespread use of Teflon, or perhaps 3M's Scotchgard fabric protector. On the phone, Crawford "pled ignorance." But he immediately suspected two 3M chemicals: PFOA or PFOS. His memo, which was circulated to company executives and to its research and medical departments, recommended "animal experiments to see just how much of these materials can, in fact, be tolerated in the bloodstream" as well as studies on the chemicals' therapeutic potential. If everybody in America was already going around with 3M's products in their blood, anyway, maybe there were "some medical possibilities that would bear looking into."[11]

By the following month, Guy and his colleagues had concluded that the mystery molecule was most likely PFOA.[12] At this point, 3M began monitoring PFOA and PFOS in the blood of its workers and found their levels were up to a thousand times higher than the average person in Guy's studies. This disparity suggested that the chemicals built up in people's bodies over time, possibly amplifying any damaging effects.[13]

To get a better handle on the implications, in 1978 3M hired an outside lab to test the effects of the two chemicals on monkeys, which are biologically more similar to people than lab rats. To the alarm of company insiders, one of the PFOS studies had to be aborted two months early because all the monkeys had died. Necropsies showed that both chemicals caused liver problems and digestive tract damage. Monkeys exposed to PFOA also developed blood and kidney abnormalities. And even those fed the lowest doses had lesions in their bone marrow, lymph nodes, and spleen—evidence that the chemical was attacking their immune systems.[14] 3M didn't disclose these findings to regulators. But it did inform DuPont, which also began tracking PFOA levels in its workers' blood.[15]

In June 1979, a group of 3M executives traveled to San Francisco for a confidential meeting with Harold Hodge, who was still engaged in bomb-project medical research. The 3M delegation walked Hodge through the findings from the monkey study and blood monitoring. According to meeting records, 3M had recently documented a rise in PFOA levels among workers and correlated it with markers of kidney and liver damage, the implication being that the chemical was causing some of the same problems in people as it had in monkeys. Given the chemicals' prevalence among the general public, Hodge advised 3M to study the effects of PFOA and PFOS in people as well as the substances' potential to cause cancer, birth defects, and other problems in animals. A week after the gathering, he called 3M and asked that a note be added to the meeting minutes. It stressed that human studies—particularly on the line of 3M products made with PFOS and used in greaseproof food packaging—were of "utmost importance." If these compounds or their "metabolites" were found to be widespread in people's bodies and to build up over time, he warned, "we could have a serious problem."[16]

3M immediately performed the birth-defects studies Hodge recommended and discovered that rats whose mothers were fed PFOA during pregnancy developed eye deformities. The company shared its findings with the EPA and with DuPont.[17] Just after Bucky's first surgery in early 1981, DuPont sent a pathologist and a birth-defects expert to review the 3M data. The pair concluded that the study—which Sue Bailey would read about on the locker room bench—was "valid" and that "the observed fetal eye defects were due to C8." At this point, DuPont pulled female workers from areas where they might come into contact with PFOA.[18]

Presumably to limit workers' exposure, the on-site lab at the Parkersburg plant designated one person per shift to analyze the level of PFOA in its products. Kenton Wamsley, a lab technician who had worked at the factory since the 1960s, recalled the day his supervisor assigned him this task: "I had an inkling that something was wrong. But he said, 'Ken, this stuff won't hurt the men.' I wasn't about to go against the paycheck that supported my family, so I shut my mouth." Before long, he developed crippling stomach pain and anal bleeding so severe that he sometimes woke up with his underwear soaked in blood.

Meanwhile, DuPont's medical director, Bruce Karrh, began secretly monitoring fifty female employees who had been exposed to PFOA. Under the guise of routine medical checks, he had his subordinates collect blood samples and get the women to fill out questionnaires inquiring about their menstrual cycles and pregnancies. The goal, Karrh explained in an internal April 1981 memo, was to "answer a single question—does PFOA cause abnormal children?"[19] The results were stark. Two of the seven pregnant workers exposed to the chemical—Bailey (referred to as "Employee W") and Robinson ("Employee X")—had given birth to babies with eye and nostril deformities similar to those found in rats. The researchers concluded that this was a "statistically significant excess" over the birth-defects rate in the general population, which was only two in a thousand.[20]

This was exactly the kind of "significant risk" data that DuPont was supposed to share with the EPA under the Toxic Substances Control Act. Instead, it simply abandoned the pregnancy study.[21] For its part, 3M collected new data refuting its own earlier research. In 1982, the company sent the EPA a paper purportedly finding no link between PFOA and deformities in rats.[22] In a meeting with agency officials, DuPont and 3M claimed that the problems 3M had previously reported were caused by researchers mangling fetal eye tissue during dissection. Meeting records show that some of the EPA officials were skeptical that "highly positive findings" could "subsequently turn out to be negative," but, as is often the case, they were reliant on industry data. The DuPont executives made no mention of the birth defects in the babies of its workers. Around the time of the meeting, the company moved women of childbearing age back into areas where they were exposed to PFOA.[23]

Although Hodge had suggested several other studies, neither 3M nor DuPont immediately pursued them. Philippe Grandjean, an expert in environmental medicine who teaches at Harvard's T. H. Chan School of Public Health and has investigated 3M's research program, found this was part of a broader pattern: The company often avoided or slow-walked studies that might find lucrative products harmful.[24] "They would rather not know than have evidence of toxicity on their hands," Grandjean explained. However, both companies continued investigating what happened to PFOA and PFOS once the chemicals left their factories. In 1984, DuPont dispatched employees to secretly fill jugs of water at gas stations and general stores around Parkersburg and bring them in for testing. Sure enough, PFOA was polluting the water supplies of two nearby towns—Lubeck, West Virginia, and Little Hocking, Ohio, just across the river from the plant.[25]

Alarmed, Karrh urged company leadership to take "all available practical steps" to reduce the public's exposure.[26] But the medical director's pleas went unheeded. That May, a group of executives gathered at company headquarters. Meeting minutes show that they weighed a proposal for cutting emissions, including adding scrubbers to vents that released PFOA into the air, but they decided not to adopt it. The additional expense was not "justified," the group concluded, since the company was already legally liable for harms caused by the PFOA it had spread over the past thirty-two years.[27]

Nothing in the minutes indicated there was any consideration given to the human toll. Nor was there any mention of the disturbing new research from 3M: While attempting to study how PFOA broke down in the environment, company scientists wound up discovering that it didn't break down *at all*—meaning every molecule of the chemical that it produced would linger on the planet for millennia.[28]

8

The Tipping Point

By early 2015, Michael Hickey was on the verge of giving up. More than a year of sleepless nights and constant worrying about the water had left him utterly exhausted. Most days, he was too tired and distracted to do anything with his son, Oliver. The sweet, rambunctious four-year-old would race around the living room trying to get his father's attention by pelting him with a Nerf gun. Michael barely looked up from his tablet. And with every passing day, Angela grew more frustrated.

Michael couldn't really blame her. He was sacrificing his family, and for what? His community was no closer to having clean drinking water. Although a handful of people had begun asking him about the situation when they ran into him at the grocery store or restaurants, most didn't seem to care. If he kept pushing, he worried he'd bring the ire of the village down on his family without making any tangible difference in people's lives. Before abandoning the cause, though, Michael decided to consult an environmental lawyer, and that January, he placed a call to a man named Dave Engel.

Engel, a scrappy, garrulous sixty-four-year-old, had spent much of his career representing corporations, but he had recently made headlines for his work on behalf of Halfmoon, New York. For nearly six years, the town had been litigating with General Electric over drinking water polluted with toxic PCBs. During that time, the attorneys representing other affected Hudson River communities had all settled, but Engel kept working the case and sparring with GE in the media.[1]

As it happened, Engel had heard about the Hoosick Falls situation a couple of weeks earlier and had been reading up on PFOA. He was amazed by the strength of the data tying it to serious diseases. On a more personal level, he was troubled to learn that PFOA was a key ingredient in ski wax.[2] His son, a die-hard skier who tuned his own skis, had a crippling case of ulcerative colitis, one of the diseases cited in the West Virginia study. The lawyer told Michael he was eager to learn more.

One gray evening, as a blizzard descended on upstate New York, Michael and Marcus traveled to Albany to meet Engel in person. Marcus was undergoing a special screening for intestinal malignancies that involved swallowing a radioactive capsule, and arrived with a monitoring device strapped to his emaciated body. He and Michael walked Engel through the details of Ersel's death, the drinking water problem, and Marcus's diagnosis.

Although Engel was moved by their story, he wasn't sure how to help. Michael and Marcus still wanted to work with, rather than around, Mayor Borge. This ruled out standard pressure tactics like leaking to the media or filing a full-blown lawsuit. Eventually, Engel suggested that Sue Hickey file a claim under New York's workers' compensation law, which entitles people to collect a portion of their deceased spouse's wages after a work-related death. Besides helping to secure her retirement, this approach would serve as a litmus test for Saint-Gobain's willingness to negotiate.

Michael balked. As he informed Engel, his family had a reputation for being litigious. In the 1990s, Sue had been watching one of the kids' Little League games when a fly ball smashed her in the back, cracking a vertebra and breaking two ribs. She and Ersel ended up suing Little League for medical expenses, and despite Ersel's standing in the community, the incident still colored people's perceptions of the family. Michael worried that if Sue filed a claim, other locals would think he'd only taken up the water issue for financial gain.[3] But the statute of limitations was about to expire, and he'd exhausted all the other options he could think of.

In the end, Michael enlisted Sue, and on the second anniversary of Ersel's passing, February 25, 2015, they filed a workers' comp claim alleging his death had been caused by "long term exposure" to PFOA.[4] Michael and Engel expected this to be the start of a long negotiation. After all, they had no blood tests to prove that Ersel had PFOA in his body, much less that it had made him sick. But a few weeks later, Saint-Gobain came forward with a confidential settlement offer, and any doubts Michael had about continuing the fight evaporated.[5] "To my mind," he said, "they were admitting they were guilty."

EARLIER THAT WINTER, Marcus had traveled to Memorial Sloan Kettering Cancer Center in Manhattan for an appointment with his surgeon. Since the chemo seemed to be working, he expected they would go over the plan for the first procedure on his liver. But the hospital rules required an oncologist to review the case first. Marcus panicked. He knew that nothing in the medical literature pointed to surgery for a person with his lung cancer diagnosis.

A few days later, the oncologist called with some startling news: Sloan Kettering had reread Marcus's biopsy and concluded he actually had "atypical carcinoid," a variation of the disease that he'd suspected from the beginning—meaning his cancer was more likely to be cured through surgery.

This news sent Michael into sleuthing mode. He ended up

unearthing a 2007 report about a carcinoid cluster among DuPont workers. There were nearly a hundred suspected cases of the rare disease. A third of the confirmed ones were diagnosed at the Parkersburg plant, which meant workers there were developing the disease at thirty times the rate in the general population.[6] Since Parkersburg was the nexus of Teflon manufacturing, he and Marcus immediately began thinking about ties to PFOA.

Marcus personally knew of at least two other area residents who had recently been diagnosed with atypical carcinoid, which suggested there was something other than bad luck at play. "We're talking about a cancer that strikes fewer than one in a hundred thousand people," he said. "Even if you combine our population with Bennington's, you're looking at maybe twenty thousand people—yet we've got three cases inside of a year. It's hard for me to believe that's a coincidence."

Marcus started drawing more connections to people in the area whose lives had been devastated by diseases associated with PFOA. Young men struggling with testicular cancer, babies born with thyroid disease. The secretary in his family's practice had lost her husband to kidney cancer at age fifty-three—and then her son to ulcerative colitis when he was just forty-two. Marcus was especially struck by the research linking PFOA to kidney problems in children.[7]

When he was five, Marcus had been diagnosed with a kidney condition called hydronephrosis. At the time, most children with cases as serious as his died. But his parents happened to know of a pediatric surgeon in Boston who had developed a groundbreaking procedure to fix urinary tract defects, and they sent Marcus there for treatment.

During the months Marcus spent in the hospital, he shared a room with three other children—a baby with brain cancer and two older boys, one of whom had stomach cancer. Decades later, he could still recall the sound of them retching in their beds after chemotherapy. The baby and one of the other boys died during his hospital stay. The one with stomach cancer seemed to improve and was released the same day as Marcus, but a few months later, Marcus learned that he had died, too. It was then that Marcus decided to go into medicine. "I want to make people better, so they don't all die like my friends died in that hospital," he told his mother, Gloria.

From that point on, Marcus devoted himself to the family's practice. In the afternoons, while other kids were out playing, the elementary schooler was in the office, scrubbing toilets and mopping floors. Tickled, Old Doc hung a brass plaque in the waiting room that read "Janitor: Marcus Martinez." Eventually, Marcus went away to college and then to medical school in Albany, but he continued working at the practice during summer breaks. In the decades since moving back to Hoosick Falls, he'd poured himself into his work. Things hadn't always gone his way in other areas. His marriage was rocky, and though he had a stepson he adored, he'd never had children of his own. Still, he had few regrets; tending to his patients, above all, gave his life purpose.

Now even that might be taken from him. Although the new diagnosis offered some hope, Marcus knew he might never recover or be able to work again. He agonized over whether to shut down his practice. If he did, he worried that other health-care providers in the area would struggle to absorb his thousands of patients, making it difficult for people to get the care they needed. But to keep the office running, he would have to pay a nearby medical center to staff it with doctors, likely putting him hundreds of thousands of dollars in debt. Marcus decided to stay open anyway. Locals did their best to return the favor. In the run-up to his surgeries, visitors showered him with cards and food, or came to thank him for the care he'd shown their families. The entire high school basketball team shaved their heads in solidarity with their beloved team doctor.

That February, Marcus traveled to Sloan Kettering, where the surgeon cut open his abdomen, carved out almost half of his liver, then worked through his intestines inch by inch with his fingers, searching for small tumors the scans had missed. He didn't find any, and Marcus bounced back fast. Hours after the surgery, he was sitting up in bed, an IV line snaking into a port in his chest as he scrolled through patient records on his phone. The following month, Marcus returned for a second surgery to remove the tumor in his lung. This time, the surgeon had to slice open Marcus's chest and cut through the diaphragm and healthy lung tissue to get at the tumor. He managed to remove it all, but the aftermath was excruciating. For a week or two after returning home, Marcus barely left the giant leather armchair in his living room, even to go to bed, because it hurt too much to lie down.

And yet by spring, he was back at work and busier than ever. As he made his way through the backlog of patients who had refused to see other doctors in his absence, he found that some of them had undiagnosed diseases that had progressed so far they were difficult to treat. At the same time, Marcus began to merge his practice with a larger one so there would be providers to see his patients when he got sick again. "The statistics for my cancer were such that I knew this was going to come back," he explained. "It was just really a matter of when."

MEANWHILE, THE PEOPLE OF HOOSICK FALLS were still drinking polluted water. Five months into its negotiation with Mayor Borge, Saint-Gobain was still refusing to provide bottled water or fund a special filtration system to remove PFOA from the public supply. Frustrated, Borge asked the village's engineering consultants to approach government agencies for help with the multimillion-dollar costs. Their appeals to state officials went nowhere, so in the spring of 2015 one of the engineers emailed a senior official in the EPA's

Office of Drinking Water, explaining that the village was "in dire need of any and all financial support and technical assistance." But the agency had no assistance to offer.[8]

The lack of progress infuriated Michael and Marcus. Since Borge was still determined not to put legal pressure on Saint-Gobain, the two men began lobbying individual board members to take action. "I know they don't want to admit this but we have a major problem with our water and it needs to be addressed ASAP . . . PLEASE!!!" Marcus pleaded in an email to one trustee.[9] Marcus and Dave Engel also enlisted a group of prominent residents and formed a nonprofit called Healthy Hoosick Water. Working closely with Michael—whose name was intentionally left off the roster because of the workers' comp claim—the group began building a case for aggressive action. They consulted environmental experts and lawyers who had handled PFOA-related cases, and tracked down former McCaffrey Street workers with information about dodgy practices like illicit dumping.

Drawing on this research, Healthy Hoosick Water hammered village representatives with letters demanding they seek an ambitious list of concessions from the company. The group wanted a robust central filtration system that Saint-Gobain would pay to maintain and operate and individual filtration for private wells, along with a program to clean up dumping sites and collect "fugitive" materials like the Teflon barrels scattered around town. They also demanded a company-funded program to monitor residents for associated diseases and, eventually, an alternate water supply to replace the village's contaminated well field.[10]

It was an audacious proposal, one with little precedent in the history of environmental cleanup. Marcus saw this approach as the only way to safeguard the community's health in the long run. But Borge—who suspected Engel was just trying to provoke a fight to generate legal fees—initially refused to even meet with Healthy

Hoosick Water to discuss the plan. So, in mid-October 2015, the organization threatened to sue Saint-Gobain unless it was given a seat at the table.[11]

Engel also placed a call to the EPA's regional office. The agency had already been alerted to the situation twice before. But in this instance, Engel spoke directly to regional chief Judith Enck, a former environmental activist appointed by Barack Obama. While most EPA officials were cautious about interfering in drinking water cleanup efforts (a process that by law the EPA is supposed to delegate to states), Enck was known for her bullish approach to enforcement. She relentlessly prodded state regulators to be more assertive and publicly blasted polluters who failed to hold up their end of the bargain.[12] She also had a personal connection to Hoosick Falls, having lived much of her life at the end of a dirt road about twenty-five miles southwest of the village.

When Enck began pressing New York officials for information, alarms apparently went off inside Saint-Gobain. The company had recently completed the site investigation it had begun during its early negotiations with Borge. It found that PFOA pollution around the McCaffrey Street plant was far worse than anyone had anticipated. Virtually all the water and soil in the area was saturated with the chemical, and the groundwater directly beneath the plant contained some of highest levels ever detected anywhere at the time. Worse, this toxic plume seemed to be migrating toward the village wells, leading to a steep spike in the PFOA levels in the treated drinking water.[13]

If the EPA got involved, Saint-Gobain could be forced to shell out millions of dollars for a years-long investigation and cleanup effort. Rather than take that risk, on October 21, the company summoned Borge and Deputy Mayor Ric DiDonato to McCaffrey Street and offered to fund a filtration system to remove PFOA from the public water supply.[14] The proposal didn't cover any of the other measures sought by Healthy Hoosick Water, and it required the village to give up all future claims involving PFOA in its drinking water. But to Borge's mind, the deal offered one key advantage: It would fix the village's most pressing problem without the devastating stigma that often came with a lengthy public process.

As he scrambled to finalize the agreement, Borge hired a law firm based in Glens Falls, New York, to advise him. However, the firm didn't take the customary measures for seeking compensation from polluters: No experts were brought on to assess the extent of the contamination or to determine whether the agreement adequately addressed it. Instead, the firm's primary focus appeared to be fending off Healthy Hoosick Water; the lawyers logged dozens of hours meeting and corresponding with Engel, who hoped to persuade them to fight for the concessions sought by Healthy Hoosick Water.[15] He was stonewalled at every turn.

But Engel was no longer the only voice demanding aggressive action. As soon as Enck, the regional EPA chief, reviewed the village's drinking water data, she switched into crisis mode. She appointed a twenty-person task force and arranged for Michael and Marcus to brief them. She also pressed senior state officials to warn residents that the water wasn't safe to drink.[16] When state authorities resisted, saying they didn't want to alarm people or disrupt the mayor's negotiations with Saint-Gobain, Enck took matters into her own hands. On November 25, she sent Borge a letter urging him to take residents off the public supply and provide an alternate source of drinking water. Confoundingly, the letter also noted that the EPA had no "funding stream" to assist with these efforts.[17] After nearly a year of pleading for help, Borge found this maddening. "Suddenly, the agency was saying, 'Don't drink the water.' But what's the alternative?" he demanded. "We were looking for alternatives and there *were none*."

In the end, Borge chose to disregard Enck's letter and circulate a misleading fact sheet from the state Health Department, saying "health effects are not expected to occur from normal use" of municipal drinking water.[18] At this point, Michael and Marcus decided it was to time to expose the entire debacle. In doing so, they would draw inspiration from a monumental struggle that had been unfolding over decades in the epicenter of Teflon production: Parkersburg, West Virginia.

9

Welcome to Beautiful Parkersburg, West Virginia

About a dozen miles down the road from DuPont's Parkersburg plant, past the clumps of crab apple and the caved-in hogpen, sat the old clapboard farmhouse that Jim Tennant called the "home place."

Jim's family had lived in the home place and farmed the fertile soil around it for more than a century. There had been some tough stretches. In the 1950s, Jim's father ran off, leaving his mother to look after nine cows, two mules, one hog, and five children. Still, the family got by, eating turtle and muskrat and peddling anything they could grow or forage—wild watercress and elderberries in the spring; ginseng and lima beans in the summer; hay and apples in the fall.

When Jim and his brother Wilbur were grown, they took over the farm. It eventually grew into a seven-hundred-acre operation, with more than two hundred head of cattle and enough corn to pack a thirty-five-foot silo. Along the way, Jim got married to a local girl named Della, and they bought a house on a plot adjoining the family's land, swapping the outhouse for an indoor toilet.

Then, around 1980, DuPont approached the couple about buying some acreage abutting their home for a landfill. At first, Jim and Della balked at having a dump so close to the farm that had sustained their family for generations. But Jim had been having health problems, including unexplained fainting spells, and they needed money for medical bills. Plus, the company assured them it would only dispose of nontoxic material, like ash and scrap metal, and so in 1983 they agreed to sell.

Shortly after closing the deal, Jim and Della noticed that their two young daughters were wheezing and hacking. Worried about the girls' health, they moved to a house in town, but most of their relatives stayed, and Jim and Della continued hunting game and eating beef grazed on the farm. Della sometimes took her daughters' Girl Scout troop there to catch tadpoles in the creek and make plaster molds of deer tracks—at least until the 1990s, when the water in the creek turned black and foamy, and the tadpoles disappeared.

On their walks around the farm, the family began finding dead deer tangled in the brambles. Their cattle started going blind, sprouting tumors, vomiting blood. One day, Jim and Della saw a cow staggering down the road in pure agony. "It was bellowing, the awfullest bellow you ever heard," Della recalled. "And every time it would bellow, blood would gush from its mouth and its nose. It just bellowed and bellowed, and blood just kept flying, and then it would fall down, and it would try to get up." Since the couple didn't have anything to shoot it with, all they could do was watch it bleed to death.

Soon the cow carcasses were piling up faster than the Tennants could bury them, and family members were being hospitalized with breathing problems and mysterious chemical burns. Convinced that the landfill was to blame, the Tennants begged regulatory agencies to investigate. The West Virginia Department of Environmental Protection eventually dispatched inspectors, who documented "numerous deficiencies" in the landfill's operations, including erosion "gullies" that funneled waste into Dry Run Creek, which ran through the Tennants' pastureland.[1] Rather than cracking down, however, in 1996 the agency struck a deal with DuPont: The company would pay a $250,000 fine, and state regulators would take no further action against the landfill, regardless of the damage to neighboring properties.[2]

But the Tennants' animals kept dying, and by the late 1990s, the EPA was asking questions. In response, DuPont proposed a collaborative investigation involving six veterinarians, half of them appointed by the company. After inspecting the family's cattle, the team concluded that the problems were caused by "deficiencies in herd management"—meaning the Tennants were to blame.[3]

Furious, the family resolved to sue DuPont. They couldn't find a local lawyer willing to go up against the company that powered so much of the region's economy, so one of their neighbors suggested they call her acquaintance, an Ohio-based environmental attorney named Robert Bilott. He wasn't exactly an obvious choice. The mild-mannered thirty-three-year-old was about to make partner at a corporate firm where he had spent much of his career representing chemical companies—sometimes against people like the Tennants. But his grandmother had lived near Parkersburg and was friendly with the Tennants' neighbors. As luck would have it, one of Bilott's fondest childhood memories was a weekend spent riding horses and milking cows on the neighbor family's farm.

Bilott agreed to meet the Tennants, and in October 1998, Wilbur arrived at the lawyer's downtown Cincinnati office, lugging several cardboard boxes stuffed with photos and papers. Seated in an airy conference room, the farmer rummaged through his cargo, pulling out snapshots of ailing animals and telling Bilott about the carnage on his family's farm.[4] Later, he put a video cassette in the VCR, and shaky footage of Dry Run Creek flickered onto the screen. Brown, frothy water. Thick, meringue-like foam bobbing in the eddies. Moribund fish floundering along one bank. The lens zoomed in on a dead deer, its nose and mouth caked with blood. "I've taken two dead deer and two dead cattle off of this ripple right here," Wilbur told the camera in his West Virginia drawl. "They're trying to cover this stuff up, but it's not going to be covered up, because I'm going to bring it out in the open for people to see." The video cut to a wild-eyed heifer with thick white slime oozing from its mouth, and then to Wilbur dissecting the same cow's carcass and pointing out its blackened teeth and bright green organs.

Bilott had planned on referring the Tennants to a colleague, but he was so troubled by what he saw that he agreed to take the case himself. By 1999, the lawyer had brought suit in a West Virginia federal court and was on the hunt for evidence. He filed motions and dug through boxes of regulatory filings on known hazardous substances being dumped at Dry Run, hoping to find something specific that could explain the blight on the Tennants' farm. None of it seemed relevant to their situation. And as news of the Tennants' lawsuit spread, other locals began treating them like lepers. Friends avoided them on the streets. Local pharmacists refused to fill their prescriptions. Sometimes, when they walked into a restaurant, all the other diners would get up and leave.

THE TENNANTS DIDN'T KNOW IT, but they had brought their case at a perilous moment for the chemical industry. For one thing, an emerging body of research was upending scientific understanding of the way chemicals affected people and the natural world. This field of study could be traced back to the late 1980s, when Theo Colborn, a sixty-year-old grandmother who had recently earned her PhD in zoology, began investigating mysterious health problems plaguing wildlife around the Great Lakes. Working with the Washington, D.C.–based Conservation Foundation,* she had collected

* The organization has since merged with the World Wildlife Fund.

scientific papers to search for patterns. Before long, her tiny office was stacked floor to ceiling with cardboard boxes full of studies detailing a bewildering array of maladies. Cancer. Immune problems. Infertility. Shrunken sexual organs. Some birds were being born with twisted beaks, missing eyes, or organs on the outside of their bodies. Others suffered from a bizarre syndrome that caused seemingly healthy chicks to waste away and die.

The sheer number and diversity of symptoms had baffled other scientists, but Colborn noticed that they had two things in common: The young were hardest hit. And in one way or another, all the animals' symptoms were linked to the endocrine system, a network of glands that controls growth, metabolism, and brain function, using hormones as its chemical messengers. This system also plays a key role in fetal development. Colborn eventually concluded that hormone-altering chemicals from everyday products like plastics and pesticides were permeating the water and causing subtle changes to developing animals' brains and organs, which could lead to grave problems later in life.[5]

Colborn was not the first to observe that such chemicals had unusual biological potency: Rachel Carson and Wilhelm Hueper had done so decades earlier. But it was Colborn who devised the theory of *endocrine disruption,* which held that these substances could affect virtually every bodily system, causing widely varied diseases and wreaking havoc on entire ecosystems. These ideas were initially greeted with skepticism by her colleagues.[6] Over the next decade, however, the zoologist built her case, collecting data and tissue samples from wildlife populations around the world and consulting ecologists studying a diverse range of habitats. One team, working around Florida's Lake Apopka, had found that the penises of young alligators were shrinking, and only one in five alligator eggs was hatching. Another group in England had discovered that male fish were sprouting eggs inside their testicles, an anomaly later attributed to the ubiquity of hormone-altering chemicals in nature.[7]

Scientists in other fields were connecting these substances to reproductive problems in people. In 1992, Danish pediatric endocrinologist Niels Skakkebaek published the first in a series of studies on the dramatic drop in sperm counts among men in Western countries. (Subsequent research has found a nearly 60 percent average decline since the 1970s.)[8] Skakkebaek suspected this trend was tied to the rapidly rising rates of male infertility, as well as the sharp increases in testicular cancer and genital deformities that he and other scientists had documented among young men. These disparate problems were, he argued, part of a broader syndrome caused by widespread exposure to hormone-altering chemicals.[9]

Laboratory research seemed to confirm his thesis. Some of the most startling findings came from the University of Missouri, where a team of researchers was studying the chemical bisphenol A, or BPA, which was used in hard plastics—including most baby bottles available at the time. The group found that male mice whose mothers were given tiny doses of BPA during pregnancy had enlarged prostates and low sperm counts.[10] Even minute quantities, it seemed, were enough to cause the kinds of disturbing health problems that Colborn had found in wildlife. Soon other scientists were linking BPA to numerous ailments—cancer, genital deformities, diabetes, heart disease, obesity, diminished IQ, and ADHD.

At the same time, it was becoming increasingly clear that hormone-disrupting chemicals didn't follow the same pattern as ordinary poisons, with symptoms increasing gradually with the level of exposure. As the University of Missouri findings suggested, even minuscule doses introduced during sensitive phases of development could lead to dire health problems down the line. They could also alter the expression of genes, giving rise to diseases that were passed down over generations.[11]

By 1996, when Colborn copublished *Our Stolen Future,* a bestselling book detailing her findings, her theory had gained widespread acceptance. That summer, Congress passed a law requiring the EPA to screen each of the roughly eighty thousand chemicals on the market for hormone-disrupting effects and begin regulating those that altered hormones.[12] Meanwhile, across the Atlantic, the newly formed European Union was embracing a novel regulatory model rooted in the so-called precautionary principle. This required *all* chemicals used in commerce—including those grandfathered in elsewhere—to be proactively tested for safety. Those that could cause cancer, disrupt hormones, or hamper fetal development were to be regulated even if the science wasn't totally settled.[13]

These policies were a major threat to the global chemical industry—many widely used chemicals had been linked to these very problems. In the case of PFOA, studies by DuPont and 3M had found that it caused tumors in the liver, pancreas, and testicles of rats.[14] DuPont suspected the testicular growths were hormone driven, in which case PFOA might be akin to a biological sleeper agent, capable of causing wildly varied diseases that manifest years after exposure.[15] In 1989, the company had conducted another study to test this theory on male rats. After just fourteen days, the rats' testicles shrank, their estrogen levels spiked, and their testosterone plunged—dramatic evidence that the chemical did, in fact, alter hormones.[16]

Within the industry, it was becoming increasingly clear that these findings were relevant to people. DuPont's routine monitoring of Parkersburg workers had found elevated rates of kidney cancer and leukemia.[17] A 1990 study involving more than three thousand employees at a 3M plant in Minnesota concluded that workers who handled PFOA for at least a decade were three times more likely than the average employee to die of prostate cancer. Significantly, the author linked even "relatively low levels" of PFOA in workers' blood to a drop in testosterone and a rise in estrogen. It also found signs that PFOA attacked the immune system, altered thyroid function, and magnified the damaging effects of alcohol and obesity on the liver.[18]

As the troubling data piled up, 3M joined forces with other corporate giants like ExxonMobil and Philip Morris that were facing possible regulation because of the growing scientific consensus about the hazards of their products. This group collaborated to establish or fund ostensibly independent scientific organizations and free-market think tanks that would prove remarkably effective at sowing doubt about the science on climate change and hormone-disrupting chemicals. They also worked to undermine science-based regulation.[19]

DuPont's leadership, meanwhile, began convening clandestine meetings where executives from around the world debated how to deal with the PFOA problem. One draft white paper from these gatherings was considered so sensitive that recipients were ordered to return it "for destruction."[20] They eventually concluded that regulation was certain to emerge somewhere in the world within five years. They began gaming out options for dealing with this threat.

One possibility was to simply "do nothing." In this scenario, the group predicted, regulators would eventually crack down on PFOA. Demand for Teflon would shrink and manufacturing costs would skyrocket. Another option was to develop a safer alternative to replace PFOA in DuPont's formulations. The group found this approach might actually drive up company profits by allowing it to seize market share from competitors, which would still be reliant on PFOA. The problem? Executives weren't certain that they'd actually be *able* to develop an alternative with the same properties that made PFOA so useful.

The team ultimately settled on a hybrid approach intended to protect DuPont's profits in almost any scenario. DuPont would move aggressively to develop PFOA replacements. At the same time, it would perform a "risk assessment"—a series of studies designed to prove that, given the right safeguards, the chemical could be used without harming human health or the environment. If the alternative chemicals didn't pan out, DuPont would use the data to stall regulators and dissuade competitors from introducing PFOA-free products.[21]

Meanwhile, DuPont's secret meetings expanded to include 3M and about a half dozen other firms that produced generic Teflon equivalents, among them *Fortune* 500 companies like Hoechst of Germany and Imperial Chemical Industries of Britain. These firms agreed to help fund costly studies that DuPont needed for its assessment, believing this work might help them dodge EU rules requiring the labeling of carcinogens and limiting toxic chemicals in food packaging.[22]

One of the program's lynchpins was a six-month study on monkeys. The earlier monkey studies by 3M had focused on signs of extreme toxicity; this one was mainly meant to track subtle effects, such as shifts in hormone and enzyme levels, from lower doses of PFOA. The goal was to prove that the mechanisms that caused cancer and other diseases in rats weren't relevant to people. But the plan backfired spectacularly. Just eleven days in, all the monkeys in the highest-dose group were so gravely ill that they had to be pulled from the study. Treatment later resumed at a reduced dose. Still, two monkeys died, one from the lowest-dose group. Autopsies revealed signs of liver toxicity, hormone disruption, and thyroid changes at every level of exposure, which suggested no level of the chemical could be presumed safe.[23]

By the time the monkey data came in, it was becoming apparent that even DuPont's most promising candidate for replacing PFOA was less effective *and* equally toxic. Meaning the company still needed the chemical for its extremely profitable Teflon line. Rather than attempt more research that could become ammunition for regulators, DuPont abandoned its risk assessment altogether.[24]

But 3M, which faced greater potential liability because it was the main manufacturer of PFOA, pressed ahead with new studies. In 1998, the company finally did what Harold Hodge had advised two decades earlier: It began investigating just how widespread fluorochemicals were in people. First, it analyzed blood samples from 645 American Red Cross donors across the country. Every single one contained PFOA, PFOS, or both. 3M then tested archived samples from past medical studies in the United States, Sweden, and remote rural China. Virtually all of them were tainted, too—the lone exception being blood collected from Korean War veterans before 1952.[25]

The findings sent shock waves through 3M, and not just because its products were apparently polluting all of humanity. Up until then, company officials had believed it was mostly PFOA accumulating in human blood. But the recent studies, based on a powerful new technology, showed this wasn't the case. The levels of PFOS, which was integral to 3M's lucrative Scotchgard brand, were higher than the those of PFOA in most samples.[26]

To get a better handle on the problem, 3M began examining how PFOS moved through the food chain and how it affected animals' development. This yielded more cause for alarm. One study found that mice pups whose mothers were fed moderate doses of PFOS during pregnancy suffered organ damage and deformities. Most died within days of birth, a sign of the chemical's devastating effects on fetal development.[27] Other research showed the chemical was likely building up to toxic levels in sea mammals and birds of prey. In one case, 3M analyzed the blood of eagles living in remote wilderness. Every animal tested, including hatchlings who had never left the nest, carried potentially harmful levels of the chemical in their bodies.[28]

This outcome troubled the study's author, a longtime 3M ecotoxicologist named Rich Purdy. In memos to company leadership, he explained that the eagle study indicated "widespread environmental contamination and food chain transfer." Substances with these properties had caused "tremendous concern within EPA, the country, and the world," Purdy noted. And PFOS was probably *more* damaging than other infamous chemicals like PCBs that built up in the body and the food chain over time, he argued, because it didn't break down. Also, rather than gravitating to soil and sediment, fluorochemicals gravitated to the blood and tissue of living beings, making them difficult to contain and remove from the environment. "I believe all this taken together constitutes a significant risk that should be reported to EPA," Purdy concluded in December 1998.[29]

3M DIDN'T TAKE PURDY'S ADVICE, but it did give the EPA some of its data on PFOS in people. The company claimed it was acting out of a sense of corporate responsibility, but there was another factor at play. In the past, detecting chemicals in liquid had been a complicated and inexact process, but by the 1990s, powerful new machines like the one 3M used for its blood studies were cropping up in university laboratories. It was just a matter of time before independent scientists began discovering PFOA and PFOS in people's blood and tracing it back to the primary manufacturer, 3M.[30]

As it braced for the EPA to make its blood data public, 3M crafted a plan to publish some of its secret studies in a way that served its own interests. According to 3M documents, its goals included placing the blood findings in a "credible context which demonstrates that there is no medical or scientific basis to attribute any adverse health effects to 3M products" and creating "defensive barriers to litigation."[31] Soon, 3M-funded papers based on long-buried research began cropping up in scientific journals, with the data massaged to downplay the more troubling findings. A late 1990s paper in the *Journal of Occupational and Environmental Medicine* falsely claimed that 3M had found "no significant hormonal changes associated with PFOA" among its workers.[32] In fact, one of the unpublished underlying studies had tied even "low levels" of exposure to changes in estrogen, testosterone, and thyroid hormone levels. Other key 3M data—particularly research suggesting that fluorochemicals were toxic at low doses or that they attacked the immune system—were kept from the public entirely.[33]

Meanwhile, 3M clamped down on internal research that didn't fit its storyline. In early 1999, Purdy, the ecotoxicologist, proposed following up on his worrisome findings in eagles with a more detailed investigation of how PFOS built up in the food chain. One of his superiors advised against this in an email, saying the research didn't align with the company's "formal plan for assessment of environmental exposure." Purdy was outraged. "Plan!" he shot back. "You continually ignore our plans and start new plans that slow the collection of data essential for our risk assessments. . . . For 20 years the division has been stalling the collection of data needed for evaluating the environmental impact of fluorochemicals."[34]

That March, Purdy issued a scathing resignation email, which branded PFOS "the most insidious pollutant" since PCBs and lambasted 3M for continuing to market this family of chemicals despite his work showing a "better than 100% probability" that one of them was building up to toxic levels in the food chain. The message, which copied senior EPA officials, also accused the company of illegally withholding "very significant" data from regulators.[35]

At this point, the EPA began pressing 3M for the buried data. And in April 2000, the company handed over a trove of internal studies, some of which alarmed the agency's upper ranks.[36] Just one month later, 3M announced that it would stop manufacturing PFOS and substances that broke down into PFOS. A company press release said the move was motivated by new data showing the chemical was present at "extremely low levels" in people and had nothing to do with safety concerns.[37] But anonymous EPA officials told *The New York Times* that if 3M hadn't halted production, the agency—which had restricted only nine chemicals in its thirty-year history—would "have taken steps to remove the product from the market."[38]

Although the release made no mention of PFOA, 3M quietly began phasing that out, too, meaning the supply stream that had sustained DuPont's Teflon brand for the last half century was about to dry up. But as usual, the Delaware giant managed to overcome this hurdle. Around 2001, it began laying plans to produce its own PFOA at a plant near Fayetteville, North Carolina, on a verdant stretch of the Cape Fear River that supplied drinking water for a quarter million people.[39]

ALL THIS TIME, Rob Bilott had been combing through the thousands of documents he'd collected, hoping to find clues about what was killing the Tennants' cattle. Finally, in August 2000, he came across a single letter that DuPont had sent the EPA in response to an inquiry about 3M products. It mentioned a chemical called PFOA being used at the Parkersburg plant, and explained that DuPont had been tracking the levels in workers' blood.* Bilott knew from experience that companies didn't do that unless they had serious safety concerns, so it instantly struck him that he might have found his culprit.

The lawyer sat for a few minutes with his heart pounding, then rushed down to the firm's library to hunt for more information. The only remotely relevant thing he could find was 3M's press release announcing the PFOS phaseout and a *New York Times* story on the same topic. So he asked DuPont to turn over any additional information in its possession.[40]

Inside the company's Delaware headquarters, his request triggered a full-blown panic. "The shit is about to hit the fan in WV," a DuPont lawyer named Bernard J. Reilly wrote in an August 2000 email. "The lawyer for the farmer finally realizes the [PFOA] issue, he is threatening to go to the press to embarrass us to pressure us to settle for big bucks. Fuck him." Reilly began coaching employees

* Specifically, the letter mentioned a type of PFOA called ammonium perfluorooctanoate, or AFPO, which appears frequently in industry documents. For the purposes of this book, the term PFOA is used to describe both variants.

to be careful about what they put in writing. "Each time you put pen to paper or fingers to keyboard and create a new document, assume you will have the plaintiffs' lawyers as recipients," he wrote in a memo to staff.[41]

After the court had ordered DuPont to turn everything over, Bilott spent hundreds of hours cross-legged on his office floor, plowing through boxes of letters, memos, and internal medical studies. Gradually, the entire horrifying story came into focus: DuPont and 3M had been studying the chemical for decades. They knew that it was toxic and that it was polluting drinking water and human blood thousands of miles away from its factories, but they had concealed most of these findings.[42]

The papers also showed that DuPont had used the landfill near the Tennants' farm as part of an increasingly elaborate cover-up. In the late 1980s, after DuPont detected PFOA in the municipal water system serving the Parkersburg suburb of Lubeck, the company set a provisional internal safety limit of 1,000 parts per trillion—roughly the equivalent of a drop of water in an Olympic-sized swimming pool. (Company lawyers reasoned that as long as pollution levels didn't exceed certain standards, DuPont wasn't obliged to inform the EPA under the provision requiring manufacturers to report data indicating a "substantial risk.")[43] To keep Lubeck's water from surging past this threshold, the company dredged up more than fourteen million pounds of PFOA-soaked sludge from the unlined pits near the community's public wells—and dumped the waste at the Dry Run landfill.[44]

Despite these dramatic measures, the PFOA levels in Lubeck's water kept climbing. The company tried to hide the problem by buying the town's well field at an above-market rate. "I feel the price difference will be justified by eliminating the use of these wells as a source of public drinking water," a manager at the Parkersburg plant wrote in a memo. "We could eliminate any future complaints or concerns about perceived or actual contamination of the aquifer by DuPont."[45]

DuPont later paid to build a new well field to supply Lubeck's drinking water and ordered employees to destroy all unanalyzed samples from the old one. It soon discovered that the new site was contaminated, too.[46] But rather than inform regulators or the public, DuPont devised a testing method that grossly underestimated PFOA levels. Reilly, the DuPont lawyer, said in an email at the time that the accuracy was "very poor" as its readings were off by "a factor of 4 or even 5."[47]

Crucially for Bilott's case, the papers also showed that DuPont had been keeping close tabs on Dry Run Creek, even as it stonewalled the Tennants. Company insiders had raised concerns about the PFOA's effect on the family's cows as early as 1991, yet DuPont had allowed ever-greater quantities to spill into Dry Run. By the mid-1990s, when the Tennants' cows started dying in droves, the levels had soared to more than eighty times DuPont's internal safety threshold. DuPont even dispatched employees to monitor the creek around the clock and dump anti-foaming agents into the water when black foam appeared.[48] (In a bleak irony, the anti-foaming agents they used to hide the pollution were themselves toxic.)

Once Bilott had pieced it all together, he invited the Tennant family up to Cincinnati and explained what was going on. All Della could think about was the countless hours her daughters and their Girl Scout troop had spent playing on the farm. She was so distraught that she had to be rushed to the hospital with heart palpitations.

Almost as soon as Bilott shared his findings with DuPont's attorneys, they agreed to settle the case for an undisclosed sum. Internal memos touted the deal as a "Win for DuPont" and a "low cost $" solution that would keep the issue out of the press.[49] But by then, Bilott couldn't imagine walking away. He had seen so many awful details in the DuPont papers that the Tennants' settlement didn't even begin to address: the studies documenting birth defects in the babies of workers like Sue Bailey; the data showing that fluorochemicals were widespread in human blood. This wasn't just about Parkersburg, he realized. It was about people everywhere.[50]

The lawyer spent the next several months drafting a letter about his findings to the EPA. With exhibits, it ran to 972 pages.[51] He also volunteered to speak at an agency hearing on 3M's PFOS phaseout. DuPont tried unsuccessfully to get a gag order against him, then scrambled to release information on the water pollution around Parkersburg before Bilott did.[52] In late October 2000, a letter written largely by DuPont officials went out on Lubeck Public Service District letterhead. It informed residents that there was a chemical called C8, or PFOA, in the water, but claimed the levels were safe to drink.[53]

10

A Rock in the Machine

One cool morning in early November 2000, a longtime Lubeck resident named Joe Kiger was relaxing on his patio swing when his wife, Darlene, handed him a letter from the water district, saying there was something called C8 in the water. At first, the message didn't strike him as particularly noteworthy. Over the next few weeks, however, he got to thinking about all the people he knew around town who had fallen ill, including three teenage boys who had recently been diagnosed with testicular cancer. His own brother, who worked at DuPont, had died of ulcerative colitis, an inflammatory bowel disease, when he was just twenty-one.

Kiger, a fifty-something PE teacher, dug the letter out of the pile on his desk and read it over and over. He kept grappling with one sentence: "DuPont has advised the District that it is confident these levels are safe." "Forget the word C8. It was DuPont that bothered me," Kiger said. "What the hell did DuPont have to do with our drinking water?"

Kiger called around to state and local agencies, pressing for more information. These requests went nowhere, but eventually he reached someone at the EPA who happened to be reviewing Bilott's 972-page letter. "I'm going to send you some information," the man told Kiger. "I want you to read it very carefully, and you'll probably want to contact a lawyer."

When Kiger went through the papers, he got sick to his stomach. Then he picked up the phone and called Rob Bilott. The timing was fortuitous: Bilott had just decided to file a class-action suit on behalf of all Parkersburg-area residents whose drinking water was polluted with PFOA. In most states, this would not have been possible—the law required would-be litigants to wait until they were sick to bring cases. But West Virginia's highest court had recently ruled that residents exposed to toxic chemicals could sue for the costs of medical monitoring to screen them for associated diseases, then seek damages retroactively if they later fell ill.[1] This gave Bilott an opening.

The lawyer asked Joe and Darlene Kiger to sign on as lead plaintiffs. Joe was willing; Darlene was wary. As a Parkersburg native, she knew how upset locals got at anyone who challenged the area's main employer—especially in an era when other manufacturers were moving jobs overseas. Sure enough, after the couple filed suit in August 2001, they were shunned by much of the community. People threw water bottles with homemade C8 labels at their house or called to insult them. One caller berated Darlene at the top of his lungs: "You're taking my job away! You're going to have to feed my kids and pay my bills if DuPont packs up and leaves because of this."

At the same time, news of the Kigers' lawsuit was helping other area residents connect the dots about the damage DuPont had done. Among them was a man named Robert Griffin, who managed the water district in Little Hocking, Ohio, a tiny village just across the river from the Parkersburg plant. After reading about Kiger's case in the local paper, he tried to get his town's water tested. DuPont, which had first detected PFOA in Little Hocking's water in 1984, refused to help. The only lab in the entire country with the capacity to test for the chemical initially declined, citing an agreement it had with DuPont.[2] When Griffin finally managed to get the testing done, he discovered that all four of Little Hocking's wells were heavily polluted.

For Sue Bailey, the moment of realization came late one evening when a private detective turned up at her door in rural Bluemont, Virginia. He told her about the Kigers' class-action suit and explained that he'd been hired by the couple's lawyers to find her as part of their investigation into a cover-up involving Teflon. "Hallelujah, prayers answered," she cried and gave him a hug.

Bailey had always believed Teflon was to blame for her son Bucky's deformities. Over the years, he'd undergone dozens of excruciating surgeries to repair them. In one case, doctors had bored a hole in the bridge of his nose, then hooked a wire through it to pull up his sagging lower eyelid. They later inserted a plastic pouch in his forehead and pumped it full of saline until he resembled a cartoon alien. That procedure left Bucky with crippling headaches, and he was so embarrassed about his bulging head that he dreaded going out in public. Using the stretched-out forehead skin—along with steel, silicon, and bits of his ear and rib—doctors were able to construct something approximating the missing half of his nose, but his face remained lopsided and misshapen.

Bucky, who was now in his twenties and optimistic by nature, had never been sure about his mother's theory. But after the detective's visit, the pair attended a community meeting about the lawsuit in Parkersburg. As he listened to stories about the strange diseases afflicting the townspeople, Bucky realized that his mother might be right. "It really felt like a punch in the face," he said.

BILOTT'S LENGTHY LETTER inspired the EPA to do some digging, too. In the fall of 2001, the agency began pushing DuPont to turn over information on the water contamination in Parkersburg.[3] Perhaps anticipating that its calculations would soon come under scrutiny, DuPont finally moved to a more accurate system for measuring PFOA in drinking water and braced for the inevitable revelation that the real levels were higher than its own safety standard. "EPA . . . better fasten their seat belts . . . We are shifting to a much better analytical method that will bring in higher numbers that may alarm citizens," Reilly, the DuPont lawyer, wrote in a pair of emails to his son, who had become his confidant and a repository for his growing frustration about DuPont's failure to protect itself from PFOA-related legal liability. "A debacle at best," he concluded. "The business did not want to deal with this issue in the 1990s, and now it is in their face, and some still are clueless."[4]

Reilly and his team began hunting for a scientific consultant who could develop a new, industry-friendly safety standard and get it "blessed" by regulators. "We need to have an independent agency agree . . . to higher levels than we have been saying, if for no other reason than we are exceeding the levels we say we set as our own guideline," he wrote.[5]

DuPont ultimately settled on a toxicologist named Michael Dourson. A small man with a trim salt-and-pepper mustache, Dourson had earned his doctorate at the University of Cincinnati's Kettering Laboratory, the department Robert Kehoe had established with money from DuPont. The two men's careers had followed strikingly similar paths. Since 1995, Dourson had run an organization called Toxicology Excellence for Risk Assessment, or TERA, which specialized in chemical-safety research. It produced and peer-reviewed scientific studies, organized conferences, and assembled panels to perform ostensibly independent risk analyses—both for corporations and for government agencies, which used them to set safety standards. An investigation coauthored by the Center for Public Integrity would later find that TERA's work had shaped "thousands of public health decisions around the country, including the setting of drinking water standards and air pollution guidelines." This despite the fact that up to 60 percent of TERA's annual funding came from industries with an interest in the group's findings.[6] Its risk assessments sometimes concluded that its funders' products were safe at levels hundreds or even thousands of times higher than the standards set by independent scientists or regulatory agencies.[7]

Not surprisingly, TERA had become a go-to for corporations looking to evade liability or keep potentially dangerous products on the market. DuPont preferred TERA, internal emails show, because Dourson had "a very good reputation" among people in "the business of blessing criteria." In other words, he could build a scientific case for an industry-friendly standard and then "sell this to EPA, or whomever we desired."[8]

In mid-2001, DuPont approached the West Virginia Department of Environmental Protection, or WVDEP, about working with TERA to set a drinking water standard for PFOA. The agency obliged, and TERA assembled a panel of scientists, half of them DuPont, 3M, or TERA employees.[9] Reilly applauded this development. "We now have established a process to come up with a 'safe' level of [PFOA] in drinking water, blessed by the agencies," he wrote in an email to colleagues.[10]

The following spring, at a public meeting at a Parkersburg high school, Dourson and the panel announced that PFOA-tainted water was safe to drink at concentrations up to 150,000 parts per trillion.[11] Bilott was so astonished that he nearly fell out of his chair. The toxicologists he'd hired for the Kigers' class-action lawsuit had found that DuPont's internal safety limit was already far too lax to protect human health. Now, TERA was proposing a standard that would allow PFOA levels in drinking water that were 150 times *higher.*[12] Because TERA's calculations were relevant to the Kigers' case, Bilott was able to question the panel's chairperson under oath. This revealed that both she and DuPont's lead toxicologist for PFOA were systematically destroying records on the panel's deliberations.[13]

Furious, Bilott appealed to the judge, who immediately ordered them to stop shredding and give Bilott any remaining papers, along with several department laptops. The salvaged records showed that the panel's early calculations would have put its safety limit on par with DuPont's internal standard, rather than the much higher number announced at the Parkersburg meeting.[14] They also revealed that WVDEP had allowed a lawyer for DuPont to edit the agency's public statements. In one case, state officials had planned to notify area residents that PFOA might be spreading through air as well as water, but the release was killed after DuPont's counsel intervened.[15]

Despite all this, West Virginia adopted the panel's recommended safety limit. And in 2003, the state tapped a lawyer who had represented DuPont in PFOA negotiations to run the WVDEP and manage enforcement of the new standard.[16]

IN SOME WAYS, the West Virginia saga is emblematic of problems that have hobbled regulation for decades. Many senior officials inside environmental agencies have spent their careers toggling between government and more lucrative corporate jobs, leading to cozy relationships between regulators and the industries they oversee. The agencies themselves often rely on industry research because they lack the resources to conduct their own.

This may explain why the EPA initially accepted TERA's recommendation, too. Under a spring 2002 agreement between the agency and West Virginia officials, DuPont was only required to provide clean drinking water to Parkersburg-area residents if the level of PFOA in their municipal supply surpassed 150,000 parts per trillion, meaning locals were stuck drinking sullied water.[17] But the information Bilott had unearthed was too damning for the agency to ignore entirely. That fall, it launched a rare "priority review" of the chemical's health effects—a sign that regulation could be looming.[18]

After combing through existing industry studies, in 2003 the EPA launched a full-scale risk assessment focusing on the potential threat to the general public. Despite industry's claims that people weren't exposed through consumer goods, the agency wound up documenting high levels of PFOA in a wide variety of items, including food packaging, dental floss, and children's clothing.[19] It also found that PFOA and other forever chemicals released from these products were permeating dust in American homes and day-care centers, where people were breathing them in.[20]

On a parallel track, the EPA began negotiating agreements with DuPont and 3M to thoroughly map the contamination around their factories and assess how PFOA from these sites wound up in people. The companies managed to limit the investigation's scope to just two of the many sites that were using the chemical—DuPont's Parkersburg plant and a 3M plant in Decatur, Georgia. They also insisted on a voluntary deal that allowed them to continue withholding some findings from the public.[21]

The results were nonetheless striking. PFOA wasn't just belching out of smokestacks and sullying the environment near factories that made or used the chemical. Huge quantities were leaching from landfills and wastewater treatment plants and polluting sewage sludge, which was then being spread over vast tracts of farmland, where it was absorbed by livestock and crops. The appalling implications weren't lost on Bilott, who repeatedly warned the EPA that PFOA was probably in drinking water supplies all across America.[22]

Manufacturers were coming to similar conclusions. When the EPA first turned its attention to PFOA, DuPont executives had begun conferring regularly with a group of companies like Saint-Gobain that used Teflon in production.[23] In 2003, the group commissioned a study, which ended up confirming that Teflon-coating operations around the country were emitting PFOA. Notably, it found the worst offenders were plants, like those in and around Hoosick Falls, that made Teflon-coated fabric. More than half of the PFOA used at these sites—thousands upon thousands of pounds—escaped through the smokestacks, infiltrating the surrounding soil and water.[24]

Up until this point, Teflon-coating operations had apparently been unaware that they were spreading more than trace amounts of PFOA, so the report sent them into crisis mode. Saint-Gobain formed a secretive multinational team of executives under the code name Tymor to coordinate damage control. The group tracked media coverage of the DuPont fiasco and assigned gatekeepers to respond if reporters began asking unwelcome questions about its own plants.[25] It also held briefings to assure key Saint-Gobain customers like McDonald's and Boeing that Teflon products were safe. Tellingly, Tymor's chairperson coached presenters at these briefings to "downplay the potential health risk of PFOA."[26]

WHAT HAD BEGUN AS A COVER-UP between two U.S. companies, DuPont and 3M, had morphed into a global conspiracy of silence involving dozens of major corporations. But their frantic maneuvering couldn't erase the facts spelled out in DuPont's internal documents, which journalists had begun to mine. In late 2003, ABC's *20/20* opened with Barbara Walters alerting her nine million viewers to the "alarming new information" about Teflon, a material used not just in pots and pans, but also in "the carpet your baby crawls on . . . your winter jacket, your skin lotion, even your makeup." The camera cut to a yellowing photo of a baby with a single nostril and a gash for an eye—Bucky Bailey. In a studio interview, Sue Bailey recounted the trauma of her son's birth: "I cried so many tears I couldn't cry another tear."[27]

At the same time, DuPont's legal woes were mounting. In 2004, a West Virginia court unsealed dozens of emails from DuPont's own lawyers blasting the business side's handling of the PFOA matter.

("God knows how they could be so clueless," one read.) The messages showed that the attorneys had been pressing the company to get Parkersburg residents clean drinking water for years—not least because doing so could radically reduce DuPont's liability.[28] The EPA, meanwhile, filed a lawsuit alleging that the chemical giant had illegally withheld key data, including studies showing PFOA crossed the placenta, potentially stunting fetal development. The U.S. Justice Department piled on with a criminal investigation on similar grounds.[29]

All the bad news sent DuPont's stock plummeting. Owing largely to concerns about Teflon, the company's share price lost nearly a third of its value.[30] Not since the furor over DuPont's alleged war profiteering in the 1930s had a PR crisis taken such a toll on its business. But the company had a plan to contain the damage. In 2003, it had reached out to the Weinberg Group, a Washington, D.C., consulting firm. Weinberg was best known for helping Big Tobacco recruit scientists to publicly argue that secondhand smoke wasn't harmful. Like other product-defense firms that came up through the tobacco wars of the 1980s and '90s, it had developed a sophisticated, multifaceted approach to seeding scientific doubt.[31]

Weinberg responded to DuPont's query with a bold five-page strategy memo. The overarching recommendation: "DUPONT MUST SHAPE THE DEBATE AT ALL LEVELS. We must implement a strategy at the outset which discourages governmental agencies, the plaintiff's bar, and misguided environmental groups from pursuing this matter any further. . . . We strive to end this now."

The memo went on to outline Weinberg's vision for a "science-based defense strategy." DuPont would deploy its lobbying might and curated industry data to "take control of the ongoing risk assessment by the EPA, looming regulatory challenges, likely [additional] litigation, and almost certain medical monitoring hurdles." Specific recommendations included hiring experts in relevant chemicals so that the plaintiffs couldn't call them as witnesses, and "constructing a study to establish not only that PFOA is safe over a range of [blood] levels, but that it offers real health benefits."[32]

DuPont followed this blueprint closely. With Weinberg's help, it recruited scientists to defend PFOA.[33] It also took another step that would prove pivotal: it hired a lobbyist named Michael McCabe. A former aide to then-Senator Joe Biden, McCabe had been chief of the EPA region that includes West Virginia when the Tennants were struggling to get regulators' attention in the 1990s. He had gone on to serve as the agency's deputy national administrator.*[34]

Drawing on his deep knowledge of the agency, McCabe presented a plan to help DuPont evade regulation. In return for certain concessions, the company would offer to voluntarily phase out PFOA. McCabe was confident the EPA would accept this arrangement, even though it wouldn't be legally binding. The Toxic Substances Control Act made it so arduous for the agency to seek binding restrictions that even a gradual phaseout might bring faster results while easing public concern.[35] In a series of private talks with agency brass, McCabe and a pair of DuPont executives laid out a list of demands. Chief among them, company emails show, were a "positive statement" from the EPA that the general public's current levels of exposure to PFOA did "not pose any adverse health effects" and a declaration that consumer products containing the chemical were "safe for use." To create a "level playing field," the DuPont team also suggested that the rest of the industry be persuaded to quit using PFOA, too.[36]

Meanwhile, in keeping with Weinberg's strategy, industry-sponsored scientists churned out studies supporting DuPont's claims about the chemical's safety. In mid-2005, a DuPont-funded paper about PFOA in consumer products appeared in the journal

---

* McCabe would later be tapped to sit on President Joe Biden's EPA transition team, which shaped the Biden administration's environmental policy.

Environmental Science & Technology. It involved many of the same items the EPA had tested, including stainproof carpet and waterproof clothing. But unlike the EPA study, it found that the levels of PFOA were negligible. "Consumer use of the articles evaluated [is] not expected to cause adverse human health effects," the authors concluded.[37] The American Council on Science and Health, or ACSH, an industry-sponsored outfit, put out a report from a group of scientists it had empaneled, apparently to counter the EPA's risk assessment. The report asserted that industry had found no evidence that PFOA made workers sick. Since the average person was exposed to lower levels than factory employees, it concluded, there was no reason to believe the chemical posed a risk at the levels of "exposure found in the general population."[38]

DuPont's PR department amplified this message with press conferences and full-page ads in newspapers across the country.[39] ACSH and a cadre of industry-funded think tanks, meanwhile, published articles portraying any attempt to regulate PFOA as just another salvo in the war on free enterprise. ACSH's president, Elizabeth Whelan, accused anti-capitalist forces of targeting Teflon precisely *because* it was so emblematic of industrial progress. "Teflon, probably more than any industrial product, is the poster child of modern technology, one that has made our lives easier and more enjoyable," she opined in an editorial for the *Washington Times*.[40]

In the end, DuPont got everything it wanted. Under the EPA's PFOA Stewardship Program, announced in January 2006, all major fluorochemical manufacturers would pledge to phase out PFOA by 2015. The deal wasn't binding, and manufacturers weren't required to disclose where the persistent chemical was being used or dumped, much less clean up the pollution. As DuPont had requested, the EPA issued a statement reassuring the public that the agency was "not aware" of any studies "relating current levels of PFOA exposure to human health effects."[41] The agency even praised the companies "for exemplifying global environmental leadership."

These assertions were wildly out of sync with the findings of the EPA's own Science Advisory Board, or SAB, an influential panel of outside experts who peer review data the agency uses to make policy. On January 30, 2006, the group issued a report finding that PFOA was a "likely" human carcinogen.[42] Two weeks later, the chair of DuPont's PFOA "Core Team," Susan Stalnecker, sent McCabe an urgent "situation analysis" of the PR fallout. "Publicity around SAB report has linked the Teflon brand to cancer. Coverage has been broad in print and network media. Significant disruptions in our markets and consumers are very, very concerned," she wrote. "In our opinion, the only voice that can cut through the negative stories, is the voice of EPA. We need EPA to quickly (like first thing tomorrow) say the following . . . Consumer products sold under the Teflon brand are safe."[43]

Once again, DuPont's lobbyist delivered. McCabe and Stalnecker arranged a private phone call between the company's CEO Charles O. Holliday and EPA chief Stephen Johnson. According to deposition testimony from DuPont executives, Holliday pressed for another broad statement affirming the safety of goods made with PFOA.[44] Almost immediately, the EPA issued one that echoed DuPont's own press releases: "The use of PFOA in the manufacturing process does not mean that people using these products would be exposed. The agency does not believe that consumers need to stop using their cookware, clothing, or other stick-resistant, stain-resistant products."[45]

WHILE IT WAS FIGHTING TO SHORE UP Teflon's public image, DuPont moved swiftly to resolve the legal cases that had exposed the brand to scrutiny in the first place. In 2005, it agreed to pay the EPA $16.5 million to settle the charges of suppressing data. This was the largest fine in the agency's history but a pittance compared to DuPont's billion-dollar-a-year Teflon profits.[46] DuPont also reached a settlement in the Kigers' class-action lawsuit. As part of the deal, the company agreed to pay $70 million in damages and install filtration systems in the area's most contaminated water districts. Moreover, DuPont committed to funding a groundbreaking community health study, juried by three independent epidemiologists who would be jointly selected by DuPont and the plaintiffs.

The rest of the deal hinged on the results of this crucial study. If the epidemiologists connected the chemical to specific diseases, locals with these ailments could bring personal injury suits against the company. DuPont would also be required to pay for a medical monitoring program to screen residents for these conditions. If the panel *didn't* find links, DuPont wasn't even obliged to keep paying for water filtration.[47] And establishing such links required much larger pools of data than could normally be collected in a single small community like Parkersburg.

This dilemma weighed heavily on Harry Deitzler, a Parkersburg-based lawyer who had worked with Bilott on the case. "I knew the reason DuPont settled the suit and agreed to assign this panel of epidemiologists was because they didn't think they were ever in this lifetime going to find links," Deitzler said. In the absence of proven associations, the plaintiffs would walk away with about $700 apiece and no further recourse. Deitzler didn't think he could face his neighbors if that was the outcome. Then one night, the solution came to him. Rather than simply divvying up the damages among the plaintiffs—a group that included all residents of contaminated Parkersburg-area water districts—they could use part of the $70 million to pay people $400 each to take part in the study, thus ensuring high rates of participation.

It was a bold idea that Deitzler wasn't sure would fly. But when he pitched it to the Kigers, Bilott, and the other lawyers over dinner, they were all for it. Deitzler knew that his fellow Appalachians wouldn't take kindly to outsiders asking probing questions about their health. So he and Bilott approached two locals—a former hospital CEO named Art Maher and a retired doctor named Paul Brooks—to handle the data collection. With the court's approval, in late 2005 the pair started a company called Brookmar, which quickly hired more than a hundred employees and built sophisticated online registration and data-tracking systems. It also placed trailers with soundproof exam rooms at four accessible locations and advertised heavily on local radio and TV.[48]

The response was overwhelming. Tens of thousands of people piled into pickups, church buses, and minivans for the pilgrimage to the trailers, where they had blood drawn and filled out a questionnaire. "We'd have families of five dragging their three kids kicking and screaming, and the parents are saying, 'Yes, you're going to get stuck in the arms—that's $2,000!' " recalled one program organizer.

By the time the trailers were hauled away in the summer of 2006, roughly 80 percent of residents in affected water districts had taken part, and Brookmar had assembled one of the largest, most detailed pools of data ever collected during a single health study.[49] This made it far more likely that epidemiologists would be able to correlate PFOA exposure with particular diseases. "I think it messed up a lot of people at DuPont's lives that we devised this wild system," said Brooks, the Brookmar cofounder. "These hillbillies threw a rock in DuPont's machine."

11

"They Poisoned the World"

Now that Brookmar had delivered, Bilott and his clients found themselves locked in a grim race against time. As part of the settlement, DuPont had agreed to fund whatever research the team of epidemiologists deemed necessary, regardless of cost. Using the wealth of Brookmar data, the panel designed eleven separate studies, including one that combined data on DuPont's emissions with sophisticated modeling of PFOA's migration through soil and water to calculate each individual plaintiff's exposure. "If somebody said, 'I was diagnosed with cancer in 1995,' this panel could actually figure out how many years prior to that the person had been exposed and what was the likely level in the water," Bilott explained. "This was something that really had never been done before."[1]

Unfortunately, this ambitious work proved time-consuming. One year passed with no results, then two, then three. Bilott did his best to keep the process moving, spending thousands of hours and vast sums of his firm's money to coordinate the analysis. But he had only so much influence, and as time went on, his original Parkersburg clients began falling ill and dying. Wilbur Tennant, who had turned up at Bilott's office with the boxes full of papers back in the 1990s, had been plagued by various diseases, including cancer, before succumbing to a heart attack in 2009. His wife died of cancer two years later, at fifty-nine. And nearly every surviving family member suffered from serious health issues. Jim, who had sold DuPont the acreage for the landfill, was dogged by heart problems. Della had thyroid disease, chronic aneurysms, and arthritis so severe that she needed ankle splints and a walker. The couple's younger daughter, whose Girl Scout troop had visited the farm, was diagnosed with breast cancer at thirty-seven and later developed thyroid cancer and gall bladder disease. (She would die in her forties.)

Bilott found it devastating that so few family members would live to see the outcome of their historic victory. Before the West Virginia farmers locked horns with DuPont, almost no one had heard of PFOA—or the broader class of chemicals it belongs to, PFAS. But thanks to the information exposed through their case, hundreds of researchers around the world were now tracking these molecules' spread through the environment. They found them virtually everywhere they looked: ringed seals in Greenland, ducks in Australia, dolphins in Brazil, household dust in China, apples in U.S. supermarkets, the breast milk of women around the globe.[2]

Environmental groups joined the hunt, too. In 2005, Tracy Carluccio of the Delaware Riverkeeper Network, a nonprofit that monitors discharges from DuPont's Chambers Works plant in New Jersey, read about the Parkersburg litigation. Curious, she began randomly knocking on doors of homes near the factory to collect tap water. Sure enough, some of the samples turned out to be heavily polluted.

Carluccio shared her findings with the United Steelworkers, which represents most of the nation's chemical workers. The union, in turn, joined forces with various regional organizations to commission water testing near DuPont plants in North Carolina, Virginia, and Ohio. All the samples contained PFOA.[3] Hoping to force a broader reckoning, United Steelworkers lobbied state officials to investigate the scale of the contamination. It also sent tens of thousands of letters to manufacturers, retailers, and fast-food chains cautioning that they had a "legal duty to warn" customers about the potentially harmful effects of PFOA in products or food packaging.[4]

As a result of these efforts, by 2006, corporate giants like Walmart and McDonald's were pledging to remove PFOA from their supply lines.[5] The New Jersey Department of Environmental Protection, or NJDEP, began a statewide analysis of twenty-three public drinking water systems, some of them in areas with no manufacturing. To the astonishment of agency scientists, nearly 80 percent of them turned out to contain PFOA, PFOS, or both.[6] NJDEP and a little-known state advisory body called the Drinking Water Quality Institute began developing standards for both chemicals—the first time any agency, state or national, had moved to regulate them.

Once again, DuPont's lobbyist stepped in. That spring, Michael McCabe and two company executives met with senior EPA officials to discuss what one internal memo described as the "activists' attacks" around DuPont's plant sites. The group pressed McCabe's former EPA colleagues to reassure the agency's regional offices about the safety of PFOA so the message "could be communicated to the public in general." McCabe and his DuPont allies also met with NJDEP scientists to advise them on methodology, hoping this would lead to laxer standards.[7] But the scientists would not be swayed, and in early 2007, the agency set a "guidance level" of 40 parts per trillion for PFOA in drinking water, twenty-five times lower than DuPont's internal safety limit. This was the first step in a process that would eventually lead to legally binding caps.[8]

One key reason New Jersey's guideline was so stringent was that it factored in new data showing that PFOA built up in people far more quickly than in lab rats. A recent study in West Virginia had found the levels in locals' blood were about one hundred times higher on average than in their drinking water, a worrying sign of how rapidly the chemical accumulated.[9] If calculations like these were applied to binding regulation, it could translate into major cleanup costs and liability for DuPont.

McCabe and his team were determined to avoid this scenario. In talks with state officials, including the governor's top economic-development aide, they stressed that DuPont was a major economic force in the Garden State. Any regulation affecting it, they threatened, could have a "direct bearing on New Jersey's economic future." They also sent a letter to Governor Jon Corzine himself, claiming the NJDEP's methodology was flawed.[10]

Suddenly, the team of NJDEP scientists working on PFOA started losing support inside their own agency. In October 2008, a group of them submitted a paper for publication explaining the logic behind the safety guideline. The NJDEP commissioner, who had been meeting with DuPont, directed the team's leader, Eileen Murphy, to pull the submission.*[11] Murphy refused. A few months later, she was demoted to a job with duties so negligible that she often filled her time by helping office assistants type letters.

Nevertheless, Murphy's former team and the Drinking Water Quality Institute pressed on. By September 2010, they were wrapping up their proposal for a binding cap on PFOA.[12] But before they could make it public, the administration of the new governor, Chris Christie, who had campaigned on promises to slash regulation, halted the process and temporarily disbanded the institute.[13] NJDEP, meanwhile, created a new industry-heavy body to advise the agency on environmental issues. Tossing aside years of painstaking analysis by the state's scientists, the group resolved to vet the safety of pollutants like PFOA using a prepackaged toxicology database—a system developed by none other than DuPont.[14]

The EPA wasn't getting any closer to meaningful regulation, ei-

* The NJDEP commissioner, Lisa Jackson, would go on to serve as national EPA chief under President Barack Obama. She declined to be interviewed for this book.

ther. Under the outgoing Bush administration, in early 2009 the EPA had issued a lax provisional "health advisory" of 400 parts per trillion for PFOA in drinking water.*[15] Enforcement-wise, it carried almost no weight; water districts didn't have to test for the chemical or inform customers if they found it, regardless of concentration. By this time, though, America was engulfed in a global financial crisis, and PFOA seemed to drop off the radar. Industry's skirmishes with labor unions and environmental groups died down; the Justice Department dropped its investigation; the EPA risk assessment stalled.[16]

All the while, communities around the country continued drinking tainted water. Just after the EPA advisory, Hoosick Falls finished work on a $7 million drinking water plant next to the McCaffrey Street factory.[17] Saint-Gobain knew enough by then to realize that the groundwater the plant drew on was probably polluted, but it didn't inform the public—nor was it required to do so under the EPA's guidelines.

BACK IN CINCINNATI, Rob Bilott was growing increasingly frustrated. His fight to draw attention to the PFOA pollution in factory towns consumed most of his waking hours. He filed lawsuits on behalf of communities in Minnesota and New Jersey and bombarded various agencies with letters laying out the problem.[18] Hoping to find a sympathetic ear in the new Obama administration, he pressed the EPA to regulate PFOA in drinking water. To underscore the urgency, he also sent officials new studies linking the substance to infertility, obesity, thyroid disease, immune suppression, and cancer.[19] But even under its more progressive leadership, the EPA

* It did so in response to concerns about water pollution in northwest Alabama, where sewage sludge polluted with industrial discharges had been spread over the land as fertilizer, leading to widespread PFAS contamination.

maintained there wasn't enough information to justify further action. And the panel of court-approved epidemiologists Bilott was counting on to fill the gaps in the data still hadn't delivered a single definitive finding.

By 2011, Bilott had begun to fear that the time and money he'd invested would never pay off for his clients, some of whom were grumbling about the glacial pace of deliberations. Even the lead plaintiff Joe Kiger had publicly accused the panel of "sashaying around the block."[20] The judge in the case was also growing impatient. That May, he summoned the parties to his chambers and delivered an ultimatum: If the epidemiologists didn't deliver soon, he would scrap the panel and have a new one appointed.[21]

Finally, around Christmas, the panel came through with its first major report. It showed a "probable link" between PFOA and pregnancy-induced high blood pressure, a life-threatening condition for both mother and fetus. By October 2012, the group had released its full findings, which Michael Hickey would stumble on during his fateful Google search after his friend's funeral. These showed "probable links" with five other conditions: kidney cancer, testicular cancer, thyroid disease, high cholesterol, and ulcerative colitis.[22]

Bilott was beside himself. After eight grueling years, a team of renowned epidemiologists had concluded that PFOA *did* make people sick. Not just workers who handled the chemical, but anyone who drank the water from their taps. Hoping this would finally convince the EPA to act, Bilott reached out to the agency once again, urging it to set strict limits for PFOA in drinking water.[23] This didn't happen, but the agency did add PFOA, PFOS, and four other forever chemicals to a list of unregulated contaminants that public water supplies must test for if serving more than ten thousand people. In early 2013, as Ersel Hickey lay dying, cities and towns around the country began their first round of sampling.[24]

• • •

ONCE THE SCIENCE PANEL HAD produced its findings, the thousands of locals with linked diseases could finally seek compensation for their suffering. Over the next few years, more than thirty-five hundred of them sued DuPont.[25] Most were longtime residents who had never heard of PFOA before the litigation. People like Kenneth Vigneron, a fifty-something truck driver who developed testicular cancer after years of drinking the water in Little Hocking. Or Sheila Lowther, a rehab nurse who noticed that an improbable number of her patients were recovering from kidney cancer surgery—and then developed a cancerous tumor in her own kidney.

Kenton Wamsley, the lab technician assigned to test PFOA in the early 1980s, was among the handful of DuPont workers who brought suit. By the time he filed, he was the only designated PFOA tester still alive. Officially, his claim cited two linked conditions—high cholesterol and ulcerative colitis—but these diagnoses didn't begin to describe the extent of his suffering. The crippling stomach cramps and anal bleeding that had plagued him during his days as a tester eventually grew so bad that he had to undergo surgery to remove scar tissue from his intestines. After that, his stomach problems eased, but he developed severe asthma, which kept him from working for long stretches. Finally, in 2001, he was diagnosed with intestinal cancer, a disease that tends to afflict ulcerative-colitis patients. His doctor informed him that he'd be dead within months unless he had his colon and anus surgically removed. Wamsley opted for surgery. This took care of the tumors, but he still suffered from asthma, fatigue, insomnia, prostate problems, chronic pain, and diarrhea so extreme that he was afraid to leave his house.

The lawsuits offered residents a way to seek a modicum of justice, but it was far from certain that they would ever see a penny in compensation. In July 2015, as the first of the personal injury cases were headed to trial, DuPont spun off its specialty-chemicals division—which made Teflon and other PFAS—into a separate firm called Chemours Company.[26] The new enterprise assumed the liability for DuPont's most polluted sites, including the Parkersburg plant, as well as its legacy of widespread PFAS contamination. But it generated only about one-fifth of DuPont's revenue—not nearly enough to cover cleanup costs, as Chemours itself would later admit.[27] After the spin-off, DuPont would merge with its rival, Dow Chemical Company, forming the world's largest chemical conglomerate. The combined company would then split into three separate firms, scattering DuPont's assets and potentially making it harder for plaintiffs to hold DuPont to account.[28]

Many people with cases pending against DuPont worried that the company would use this arrangement to avoid paying damages or drag out the process even further. "I'm sure part of their theory is the longer they delay, the more people will die," said Deitzler, the Parkersburg-based lawyer. "It's already worked. Before we could even file cases, many of the people who've been affected passed on."

DuPont appeared to be skirting liability in other ways, too. The science panel's findings had triggered the provision requiring DuPont to pay for medical monitoring to screen locals for linked conditions. Bilott saw this as the heart of the whole settlement since it would allow illnesses to be caught early, when they were most treatable. His Parkersburg-based colleagues wanted Brookmar, which had so successfully overseen data collection, to manage the medical monitoring. Instead, DuPont maneuvered to have it run by a lawyer named Michael Rozen, who had played a key role in administering the fund to settle claims from BP's Deepwater Horizon oil spill. The firm he worked for at the time had been sued by multiple Gulf Coast residents, who accused it of delaying payment for as long as possible, then offering financially desperate claimants a fraction of the money they were entitled to.[29]

Many Parkersburg plaintiffs believed Rozen was deploying a similar strategy in their community. Information sessions were held at times when most people were working, and enrollment packets were so complicated that, early on, only a few hundred residents bothered filling them out.* Brooks, the doctor behind Brookmar, was convinced that DuPont had intentionally set up the program to fail. "They poisoned the world," he said. "A successful medical monitoring program would give us much better data on the links between this chemical and various diseases." In which case, Brooks noted, "DuPont would have so much liability that it couldn't possibly compensate everyone."

After fourteen years of battling DuPont, Joe Kiger found all this maddening. He and a handful of other frustrated residents took to picketing stockholders' meetings to pressure the company to honor its commitments. Meanwhile, Kiger's own health was deteriorating. Even while he was in the hospital for heart surgery, he fielded phone calls from people with questions about PFOA and scoured his bulging satchel of papers for answers. His wife Darlene had begun to fret about the toll the struggle was taking. "He never gets a break from it," she said. "It's been so many years of watching this thing eat at him every single day, and I wonder, is it ever going to end?"

---

* Rozen strenuously denied these allegations and asserted he had done his best to encourage participation. "The benefit that is being provided to the class is exactly what was prescribed and then some, by the parties themselves in their negotiated settlement," he said. The program would ultimately enroll about thirty-five hundred of the roughly eighty thousand eligible residents.

12

The Reckoning

When the first major stories about the Parkersburg saga appeared in national media in late 2015, Michael Hickey was horrified.[1] Of course, he'd always suspected there were things that industry was hiding, but the details were so appalling. All those years, while Ersel was grinding away at the factory, Teflon splattering his arms, DuPont had known that PFOA made workers sick. It had known the chemical was saturating the water in factory towns. But instead of warning people, it had covered up its own research.[2]

Similarly, as other Hoosick Falls residents read about the devastation PFOA had caused in Parkersburg, some began to doubt the official assurances that their own water was safe to drink. Hoping to allay these concerns, Mayor Borge announced a public forum on the water for December 2, 2015. That morning, Dave Engel called Judith Enck at the EPA, worried that the mayor would use the event to spread more misinformation. Enck told him about the letter she'd sent urging Borge to take people off the municipal water—a letter that was never made public.[3] Shocked, Engel alerted Michael and

Marcus, then hightailed it to Hoosick Falls with three hundred copies of the letter in his trunk.

By the time residents began trickling down the stairs to the basement of the Immaculate Conception Church in puffy coats and baseball caps, state officials were gathered on one end of the cavernous yellow room with Mayor Borge and Saint-Gobain representatives, handing out packets saying the water was safe for "normal use."[4] At the other end of the room, standing behind a plastic table they'd hastily assembled with permission from the village priest, Michael, Marcus, and Engel distributed findings from the West Virginia study, along with Enck's letter. As people read through the materials, dismay registered on their faces. Some stormed over to Borge, demanding to know why he hadn't shared the letter. Others crowded around Marcus, yelling questions about their families' health over the din of anxious voices. Word quickly filtered back to Saint-Gobain headquarters; in response, the company's health and safety director emailed Borge with some suggestions from the PR department about how to "change the room dynamic."[5]

But Borge hadn't just lost control of the room; he'd lost control of the entire narrative. Before the forum, Michael and Marcus had been feeding information to a reporter at the region's largest paper, the Albany *Times Union*. Shortly after the event, the paper published its story, "A Danger That Lurks Below." It explained how Michael had unearthed the pollution and exposed how state and local authorities had downplayed the health risks and allowed people to keep drinking tainted water for well over a year.[6] The news sent shock waves through the village. With anxious emails pouring in, Borge finally agreed to take the village off the public water supply. He also announced that Saint-Gobain would make bottled water available at the village's lone grocery store—meaning that a full fifteen months after Michael discovered the contamination, people finally had access to clean drinking water.[7]

Through all of this, Michael's phone was buzzing nonstop. The engineer who redesigned the McCaffrey Street towers in the 1980s, enabling them to pump out three times more coated fabric, called to say how guilty he felt that his work might have added to the pollution. (He hadn't known about PFOA at the time.) Other locals reached out about their families' struggles with heart problems, thyroid disease, infertility, leukemia, kidney cancer, and testicular cancer, among other maladies. If Michael saw a possible link to PFOA, he shared any information he felt would be helpful and suggested they make an appointment with Marcus, who had developed a system to screen patients for associated diseases. Even when Michael suspected there was no connection, he listened intently to people's stories because, he figured, that's what Ersel would have done. But it was clear that many locals needed more than a sympathetic ear. They needed to know how to protect their families. So Healthy Hoosick Water began organizing its own public forum.

In the run-up to the event, the contamination crisis in Flint, Michigan, exploded into the national headlines. The parallels with Hoosick Falls were striking. Officials at every level had known since 2014 that Flint's water was tainted, but they had failed to do anything until citizens forced their hand. Now federal prosecutors were investigating, and images of the fallout were flashing across TV screens everywhere: anguished parents lugging cases of water and herding children into emergency blood-testing centers; protesters pouring into the streets with plastic bottles full of rust-colored sludge and signs that read "Not Your Lab Rats" or "Don't Poison Our Kids!"

Against this backdrop, on January 14, 2016, about a thousand Hoosick Falls residents packed into the public school auditorium. Marcus, the moderator, kicked off the discussion by recounting how he and Michael had unearthed the contamination. Judith Enck of the EPA explained the water sampling results and outlined the steps residents should take to avoid further exposure: "Do not drink the water from the Hoosick Falls public water supply. Do not cook with the water. No humidifiers. . . . Children and people with rashes should avoid long showers and long baths . . ." A toxicologist walked the audience through PFOA's effects on the body. It took nearly three years for the average person to shed just half their PFOA burden, he stressed, meaning people with high levels would carry it for decades. A hush settled over the room. But when Marcus finally gathered the presenters around a long table on the stage, people couldn't get their questions out fast enough: Was it safe to bathe infants? What about eating vegetables from their gardens? Or meat and milk from local farms? What about maple syrup? Or the public swimming pool? Or the Little League field next to Saint-Gobain? Were they polluted, too?

About two hours into the proceedings, a redheaded organic farmer named Marianne Zwicklbauer stood up. "One of the things I think as I sit here, as I'm listening, is how unbelievably let down I feel by the EPA, specifically, the fact that you all knew about PFOA in the year 2000 and all of DuPont's information about how it made people sick and how it caused birth defects," she said. "Why weren't [you] warning our community? Why did it take a concerned citizen?" she demanded, to a flurry of applause. "I thank God for the concerned citizens that brought it to this point, but I just feel so let down by my government, by you, by the fact that you had all that knowledge, and you aren't out saving these people!"

JUST AFTER THE FORUM, National Guard trucks rumbled into Flint to distribute bottled water. President Obama declared a state of emergency. Michigan's attorney general launched a criminal investigation into the mishandling of the water crisis, and federal lawmakers began demanding the resignation of Governor Rick Snyder. Protesters flooded the streets around Snyder's Ann Arbor home, while another group stormed the state capital, chanting, "Rick Snyder has got to go!"[8]

The spectacle stirred a panic in New York's state capitol. In late January, Governor Andrew Cuomo—whose administration was suddenly under intense scrutiny for its handling of the Hoosick Falls crisis—summoned Borge for an emergency meeting with health and environmental officials. Afterward, visibly shaken Cuomo aides herded Borge onto the stage for a press conference and announced a series of steps to "restore the public's confidence in Hoosick Falls."[9] These included providing blood testing for residents and setting statewide standards for PFOA and PFOS in drinking water. The officials also declared Saint-Gobain's McCaffrey Street plant a state Superfund site and announced an emergency order to classify PFOA as a hazardous substance—unprecedented moves that gave New York the power to investigate the breadth of the pollution and force those responsible to pay for the cleanup.[10]

Cuomo's reversal was a triumph for community advocates like Marcus and Michael. After nearly two years of inertia, authorities were finally acting to address the contamination, using some of the very tools that Healthy Hoosick Water had been advocating all along. But the announcement had some grim, unintended consequences for the people of Hoosick Falls. Area banks stopped giving out mortgages for homes in the village, throwing the market into chaos.[11] Some older residents who had used their homes as nest eggs had to delay moving to assisted living, while families who were in the middle of buying one house and selling another were left with no place to live. Most of the development projects that had sparked such optimism about the village's future also stalled.

Instead, the town was attracting a less flattering kind of attention. In late January, Erin Brockovich turned up at the Falls Diner wrapped in a long black coat and mingled with wide-eyed locals. Afterward, she hosted a forum with a team of personal injury lawyers to recruit plaintiffs for a class-action lawsuit.[12]

Soon attorneys from across the Eastern Seaboard were arriving in Hoosick Falls to sell their services. Reporters patrolled the downtown streets and squatted in the grocery store parking lot, waylaying residents as they lugged bottled water to their cars. Many of these journalists also knocked on Michael's door. With his dread of public speaking, Michael always felt awkward around cameras and microphones. But he didn't want angrier, less-informed people to be the ones representing Hoosick Falls—and part of him craved recognition for the work he'd done. So he assumed the role of spokesman, which invariably meant dredging up the painful story of his father's death in interview after interview.

Not everyone was grateful for Michael's efforts. With the town awash in rumors that his family had gotten a payout from Saint-Gobain (few realized it was a workers' comp settlement), some locals figured he was hyping the issue just to enrich himself. People he'd known for years began avoiding his gaze when he ran into them at restaurants and Little League games.

Things only got uglier when Michael began organizing another lawsuit—this one aimed at getting a medical monitoring program like the one in Parkersburg. This was no mean task; like many states, New York required plaintiffs to prove that they'd been injured *before* seeking damages.[13] But Michael had found an environmental lawyer named Stephen Schwarz who had a plan to overcome this hurdle. Specifically, Schwarz would argue that simply being exposed to PFOA was a form of injury, since it stayed in the body for years, disrupting key biological systems.

With all the animosity coming his way, Michael had hoped to find another local to serve as lead plaintiff on the case, which also sought compensation for those harmed by contaminated private wells or the drop in property values. But none of the people he approached were willing to take on Saint-Gobain. So, that spring, he filed a class action on behalf of every Hoosick Falls resident against the French company and Honeywell International, a U.S. firm that had operated the McCaffrey Street plant and several other area factories in the 1980s and '90s.[14] (DuPont and 3M were later added as defendants, since their products were allegedly to blame for the water contamination.)

The blowback was swift and brutal. People muttered insults when they passed Michael on the street and blasted him on social media. "Come on Hickey really you money hungry fool," read one comment. "Your poor brother Jeff he is a good guy. Why he hasn't punch[ed] you in the mouth is beyond me. Your dad, hell, I don't really think he would be all that proud of you."

Angela found these attacks crushing. After several months of scouring Facebook and Twitter for the nasty insults that sprouted every time Michael spoke publicly, she started ignoring his interviews—and tuning him out whenever he spoke about the water at home. "I separated myself, which sounds horrible, but it would kill me, it would rip me apart, to see what people would say about Michael," Angela said.

As Angela pulled away, Michael started spending less time at home and more time out drinking. Once or twice a week, he would turn up at a local dive bar in his BMW and blow hundreds of dollars on pull-ticket gambling and shots for his buddies. This didn't exactly endear him to people who thought he was in it for the money. One evening, a supervisor for Saint-Gobain saw Michael out partying and tried to jump him. The other bar patrons held him off, but Michael's factory-worker brother warned him he was going to have to watch his back. "These guys see you as threatening their livelihoods," Jeff cautioned. "You have no idea what any of them will do once they get a few drinks in them."

13

Cloud Nine

Sandy Sumner, a sixty-something woodworker from North Bennington, Vermont, was at home watching television one evening in February 2016 when a lanky man with a black satchel knocked on his door. The visitor explained that he had been sent by the state Department of Environmental Conservation, or DEC, to conduct some impromptu tests on private wells near the defunct ChemFab plant. Just a precaution, he assured Sumner. Probably nothing to worry about.

Sumner had been living in the maple-covered hills of North Bennington for about twenty-five years. When he first built his pale-green cottage there, the air smelled so fresh that he barely even noticed the factory in the valley below. But in the mid-1990s, when ChemFab ratcheted up production to meet the demand from Saudi Arabia for coated fabric, he and his neighbors noticed fumes that reeked of burning hair. Soon, they were coming down with migraines, sore throats, and nosebleeds.

Sumner and his fellow residents sought help from the EPA, but nothing changed. Then, in the early aughts, the developer of their subdivision, who'd been frustrated by dropping sales, sent away for ChemFab's permits. He discovered that one of the plant's smokestacks was operating without a scrubber, in violation of Vermont law, and confronted the DEC.[1] The agency in turn confronted ChemFab, which had recently been acquired by Saint-Gobain. Instead of tightening emissions, in 2002 the French chemical giant simply shuttered the North Bennington plant and redeployed the equipment to its facility in Merrimack, New Hampshire, where there were looser regulations.[2] (Saint-Gobain also transferred the McCaffrey Street plant's coating towers to Merrimack during this period, though it continued to produce other Teflon products in Hoosick Falls.)[3]

Once the factory had closed, the stink disappeared, and Sumner considered the problem solved. So it came as a terrible shock when, a couple of weeks after the technician's visit, he received a call saying his water was heavily contaminated with PFOA. Sumner immediately went door to door to inform his neighbors, many of whom turned out to have linked diseases—the young policeman two doors down was being treated for testicular cancer and had just had a testicle surgically removed, leaving him unable to father children.

That same day, Vermont's DEC Commissioner Alyssa Schuren was in her office when the director of the agency's waste management division burst in with urgent news about the North Bennington investigation. Not only were the wells near the plant contaminated, but a set of wells a half mile away were, too—suggesting the pollution was widespread.

Schuren rushed to the state capitol to inform Governor Peter Shumlin, who convened a meeting with his top aides and Bennington's state senators. Hoping to avoid the kind of uproar they'd witnessed in Hoosick Falls, they decided to hold a press conference right then and a public meeting in North Bennington the next day.

During the three-hour drive to North Bennington, Schuren was on and off the phone with her team in Montpelier. "Our technical staff didn't know a lot about this chemical, so we were scrambling," she said. Though it wasn't standard procedure, she decided to issue a do-not-drink order for a large part of the community and make bottled water available to people living there. The question was how large an area the order should cover. When Schuren arrived at the meeting site, she and her team were still debating even as residents and camera crews crammed in for the announcement. Finally, Schuren directed a deputy to draw a circle with a mile-and-a-half radius on the aerial map they had brought for the presentation. "Since we knew people's blood levels would increase if we had them drinking this chemical even one more day, we decided to just hit pause," she explained.

The following day, as news trucks and National Guard tankers descended on tranquil North Bennington, Saint-Gobain informed New Hampshire officials that it had quietly sampled the public water supply near its Merrimack factory and found PFOA there, too.[4] By then, New York had also detected high levels in the municipal water near the Taconic Plastics plant in Petersburgh, and both states had begun systematically testing private wells near the Teflon plants within their borders.[5] Soon, hundreds of homeowners in the area would be hit with devastating news.

EMILY MARPE CALLED HER HOME CLOUD NINE. Her mother had painted the name on a sign that hung on a stump in front of Emily's pale-yellow ranch house, near the stone patio and the flower bed bursting with bleeding hearts and black-eyed Susans. It seemed to capture the joy she felt at finally having a real home for her family.

For much of her teens and twenties, Emily had lived with her father in a dented two-bedroom trailer just off the highway south of Hoosick Falls. At seventeen, she got pregnant. She doubled up on coursework and took shifts at the local pizza joint so she could graduate from high school and save some money before her son Ethan was born. About five years later, her daughter Gwen arrived; Emily's father moved out of the trailer, and her new boyfriend, Jay, moved in. A shy, barrel-chested man, Jay worked part-time at Home Depot. Emily, who before getting pregnant had planned on becoming a nurse, took a job as an aide at an assisted living facility. She found the work gratifying. And she could squeeze forty hours into a single weekend, allowing her to spend most of her days at home with the kids, reading board books and playing with Legos.

Eventually, Jay's two older daughters started spending weekends at the trailer. For a while, the couple made it work, bathing the kids in pairs and stacking bunks in the second bedroom. But once the older children started hitting puberty, things got more complicated. Emily, who was in her late twenties then, started saving money in the hopes of one day buying a home.

A plainspoken woman with soft-gray eyes and wavy sand-colored hair, Emily didn't come across as particularly worldly. She'd never traveled far from home. Her media diet consisted mainly of romance novels and local TV. She rarely wore anything besides baggy cotton pants and T-shirts with an old fleece jacket. But beneath her unassuming demeanor was a keen intellect. She was also unusually determined: Whenever Emily set her mind to doing something, it got done. Within a year of deciding to move, she had managed to sock away $6,000—more than a fifth of her annual salary—and to qualify for a low-income USDA mortgage. Most of the places in her price range needed a good deal of work. When she first toured the Petersburgh property, the tile in the bathroom was chipped and mildewy, and the garage was piled to the ceiling with junk. But it had three bedrooms and a spacious living room with oak floors and a view of the Green Mountains.

Emily scraped together another $6,000 in family loans for the down payment and made an offer. On the day of the closing in late 2011, her mother, who had never owned a home, sat at her side and wept. Then the seller handed Emily the keys and asked her how she felt. "I'm on cloud nine!" Emily replied. Giddy, Gwen and Ethan took down the FOR SALE sign and skipped up the driveway, holding it above their heads like a trophy.

For a while, Emily and Jay poured most of their spare time and money into fixing up the house. They refinished the wood floors, gutted and rebuilt the main bathroom, and remodeled the kitchen, ripping out the grungy carpet and installing speckled gray Corian counters—a cherished gift from Emily's grandfather. As a finishing touch, Emily painted the walls soft ocean hues, stenciling on hopeful aphorisms: You are STRONGER than you seem, BRAVER than you believe, and SMARTER than you think you are.

Having a true refuge drew their close family even closer. They spent most of their free time together, playing cards, shooting hoops, or camping on the lawn. Meanwhile, the children flourished. Gwen, a bright, gregarious girl, made honor roll and joined the school basketball team. Jay and Emily eventually began setting aside money and talking about having another baby. Things had been so precarious when Ethan and Gwen were little. Now, the couple was in a position to really savor the early years.

Then, in February 2016, a technician from the county came to test the family's private well for a chemical called PFOA, which was being used at the nearby Taconic Plastics factory. Emily started reading up on the substance and then the whole sordid Parkersburg saga. That's how she realized that the same chemical that was being blamed for birth defects in babies and cancer in thousands of people was wafting from the smokestacks just down the road from Cloud Nine.

A few weeks later, the county's environmental health director called to inform Emily that the water her family had been drinking and bathing in contained 2,100 parts per trillion PFOA—five times the EPA's safety limit. Still gripping the phone, she fell to her knees, dry-heaving.

Emily started obsessively researching PFOA's health effects. To her horror, she realized that family members who lived nearby suffered from some of these conditions. Her grandmother had kidney cancer; her mother had ulcerative colitis. Gwen, who was only ten, had recently discovered lumps in her breast, which was all the more worrisome now that Emily realized that PFOA was linked to abnormal breast development and mammary cancers.[6]

Emily took Gwen to the hospital for an ultrasound, which found that the growths were most likely benign. Emily knew she should be relieved. But afterward, when she took Gwen to the local deer park, she stayed in the car, dizzy and short of breath. Then a searing pain spread between her shoulder blades. She called for Gwen to get in the car, then sped back to the hospital.

The doctor in the emergency room pegged the episode as a panic attack. He prescribed a sedative and told Emily to get some rest. Instead, almost as soon as she was home, she dove back into her research. She didn't know what else to do: Despite the furor in nearby Hoosick Falls, Petersburgh residents still hadn't been given guidance on basic questions like whether it was safe to bathe. But the morass of information Emily found on the internet only left her more anxious and confused.

Then, one evening, she stumbled on Michael Hickey's Facebook page. She knew from news reports that he was the main reason she'd found out about her water when she did. So she posted a message thanking him. "Right now," she added, "I'm alone sitting in a house I put everything I had into for my children and feel as if every minute I stay I'm killing them slowly [but] surely." Michael responded, as he usually did, by posting his phone number. A few hours later, he was on the line listening to her story.

Emily found Michael's deep knowledge soothing. Soon they were talking almost every day, and much like Michael, Emily was staying up until three or four each morning reading scientific studies. Then she'd get up at seven to go to her new job, caring for developmentally disabled adults. During her off hours, Emily emailed state agencies, circulated petitions, and rallied neighbors to attend town board meetings and demand provisions like bottled-water delivery, blood testing for residents, and a filtration system for the local elementary school. She and another woman would show up at the town's twice-weekly bottled-water distributions, sometimes standing for hours in the mud and rain to hand out fact sheets on PFOA, along with tips on bathing and cooking. Neighbors plied her with questions. "Things started clicking for people, and some of them were very scared," she said.

Eventually, Emily started co-organizing public forums that brought lawyers, scientists, and officials from various agencies to speak with residents about PFOA-related issues. It was through these gatherings that state officials finally laid out plans for cleaning up the local drinking water. By then, Emily had a firmer grasp on the science than some of these bureaucrats did, and she was quick to challenge them when they got facts wrong or offered false assurances. She also turned out to have a talent for boiling down complex details into simple language, making her a favorite among local reporters.

Unlike Michael, Emily couldn't care less what other people thought of her advocacy. One day, a town official approached her and asked whether she was trying to lose hundreds of Taconic workers their jobs. "Are you trying to kill my two children?" she shot back. But Emily hated how much her activism was taking her away from her family. She rarely made it home for dinner, much less card games and slumber parties on the lawn. "She went from being there for everything to not being there at all," Gwen recalled. And despite her dogged research, Emily wasn't any closer to answering the question that kept her awake at night—what the polluted well meant for her family's future. However, that spring she signed a retainer with Michael's lawyer and began organizing a class-action suit seeking medical monitoring for her children and other exposed Petersburgh residents. The county also started offering blood testing. Hoping this might finally shed some light, in late March Emily took her family to get their blood drawn.

BY EARLY 2016, when New York began testing private wells, the PFOA crisis had spread to dozens of communities across the Midwest and Northeast, leading to a steady stream of national headlines. Officials in affected states struggled to address the problem, given the lack of federal regulation and conflicting signals from the EPA, which was enforcing stricter standards in some communities than in others. So that March, a group of governors began pressing the agency to revisit the science and issue new guidelines, as Rob Bilott had been doing for years.[7] At the same time, federal lawmakers began demanding an investigation into the mishandling of the crisis in Hoosick Falls.[8]

Finally, in mid-May, the EPA unveiled a new "lifetime health advisory" of 70 parts per trillion for PFOA and PFOS combined in drinking water, meaning the agency only considered these chemicals safe at levels about six times lower than the guidelines it had issued during the final days of the Bush administration.[9] Although such advisories aren't binding—they don't require polluters or government agencies to supply residents with clean drinking water—they do obligate water districts to disclose the presence and potential hazards of the chemical involved. Suddenly, more than five million Americans in nineteen states and several U.S. territories were informed that their drinking water contained unsafe levels of these chemicals.[10] In dozens of communities across the country, authorities issued do-not-drink orders or shut down public wells. Panicked residents packed into public meetings and swarmed the parking lots where emergency workers handed out bottled water.

But because PFOA remained unregulated, there was little or no federal funding available to help communities deal with this situation. The only options open to most local officials were to pass the costs on to the public or negotiate with the polluter, which, in many cases, was also the largest employer in the area.

Meanwhile, in early June, blood-test results began arriving in Hoosick Falls and neighboring towns.[11] Emily came home one afternoon to find a stack of letters from the state health department in her mailbox: one for each member of the family. Unlike some people who didn't know what to make of the numbers in the envelopes, Emily had been gaming out scenarios for weeks. She knew that the average Parkersburg resident had about 70,000 parts per trillion PFOA in their blood.[12] She also knew that those people had drunk polluted water for decades, whereas her family had been on their well for only four and a half years. So she was desperately hoping their numbers would be lower than 70,000.[13]

Heart pounding, Emily sat in the car for few moments, shuffling the envelopes. Finally, she mustered the courage to open the letter with Gwen's name on it. Her number was 207,000. Struggling to breathe, Emily picked up her phone and messaged a friend: "God Gwen 207. I'm gonna have a heart attack." The other envelopes contained more bad news. Emily's blood level was 322,000 parts per trillion, and Jay's was 418,000, on par with the average worker at DuPont's Teflon factories.[14] Emily gathered the papers, rushed inside to her room, and locked the door so her children wouldn't see her cry.

14

Dirty Water, Dirty Deal

The same week that Emily got her family's blood-test results, Rob and Heather Allen of Hoosick Falls received the envelopes for themselves and their four children. The young couple had read about the contamination crisis in the papers and were anxious enough that they'd been driving two hours round trip to bathe the kids at a relative's house. But it wasn't until they opened the letters that the devastating implications truly sank in: Their sweet, fuzzy-haired two-year-old daughter, Emma, had PFOA levels that were fifty times the national average. Heather, who had breastfed her babies, spent several hours wandering around in a daze. "I couldn't get past the idea that *I* did this to my children, that it was *my* fault," she said.

Rob hated seeing her so consumed by guilt when it was clear to him that the blame lay elsewhere. After all, for most of his daughter's life, officials had been telling residents the water was safe to drink. Unable to fathom how they could have gotten it so wrong, he asked Michael Hickey to share the research he'd collected over the years. Every evening from then on, Rob would settle into the scruffy corduroy recliner in his living room, surrounded by toys and stray children's clothing, and read for hours, texting and calling Michael as he went. He scoured regulatory filings. He dug through formerly secret DuPont documents. He pored over thousands of government emails obtained via public records requests, carefully linking every official misstep with Emma's developmental milestones. When news of the PFOA contamination first reached state officials, Emma had been just three months old. She was only seven months old when the EPA was first alerted—yet it would take another year before the agency acted. During this time, Heather kept on nursing, unaware that her milk was tainted with PFOA she'd absorbed from the water.

Rob, a balding, bespectacled thirty-eight-year-old high school music teacher, had never taken much interest in politics. But now he began diligently attending public meetings and urging residents to take political action. That summer, when the state put out a promotional video touting its environmental record, Rob responded with his own elaborately produced video, featuring a score he'd composed especially for the occasion.

His goal was simply to highlight what he saw as the state's hypocrisy, but the result was a full-throated rallying cry. He urged residents to fight to protect their families and laid out specific goals, including systematic testing for chemicals like PFOA in public water supplies statewide as well as government hearings to "fully investigate why our community was allowed to be knowingly poisoned." "Join us as we lead the way in protecting the environment, protecting our citizens," he implored.

Within a week, the video had amassed thousands of views and was being picked up by regional television stations. Albany's PBS affiliate made it the centerpiece of a lengthy segment that also featured an interview with Rob and clips of tiny Emma crawling around in her sundress and diapers.[1] He and other residents who'd been galvanized by their test results also started organizing a rally at the state capitol to demand hearings and a program to monitor people for the diseases linked to the chemical in their bloodstream.

By then, many locals were channeling their anxiety into activism. Loreen Hackett, a life-long Hoosick Falls resident and heavy metal enthusiast, got a crash course in social media from her twenty-something daughter. She then set up a Twitter account, posting a series of haunting black-and-white photos she'd taken of residents with their blood levels handwritten on signs or on their own bodies.[2] She and other concerned locals also lobbied the state legislature, while a group of high school students held a press conference calling on the Cuomo administration to fund an entirely new water supply.[3]

On the day of the rally, in mid-June 2016, these varied factions all descended on the state capitol. Taking their cue from Loreen's images, they had their families' blood levels scrawled on their bodies or written on signs hung around their necks.[4] One by one, they addressed the assembled reporters. "My mom came to Hoosick Falls for a good reason, but now she knows the whole time she was trying to raise us we were being poisoned," a ten-year-old girl named Hailey Bussey declared as the crowd chanted, "Hold the Hearings! Hold the Hearings!"

But not everyone was being so civil. Livid that authorities had waited so long to inform them about the water, some people were bashing state and local officials on social media. "These idiots [are] putting our children's lives at risk for political gain!" read one comment. "Punch someone in the face that's in charge," read another. "Maybe then they will start to care." Deputy Mayor Ric DiDonato's teenage son was being bullied at school, and Mayor Borge was getting death threats.

The tensions were palpable in early July, when Senator Kirsten Gillibrand held a public roundtable with area residents and called for both state and federal hearings into the crisis. Michael, who was sitting near Gillibrand on stage, was troubled by her proposal. Although he was heartened to see politicians finally paying attention to Hoosick Falls, he was afraid that focusing on past mistakes would only deepen the rifts in his community.

Gathering his courage, he stood up and stammered something about hearings not being necessary. This infuriated some people in the audience. A few shouted insults or stomped out of the auditorium. "He might have gotten whatever he needs out of this, but the rest of this community hasn't," one woman seethed to a television reporter on her way out the door. "They're saying he's some kind of hero," another local grumbled. "He kept his mouth shut for eighteen months. . . . He's just as bad as them."[5]

The truth was that nothing Michael or his fellow residents said could have altered the outcome: The scandal was now too big to ignore. That same day, Congress opened an investigation into the official response to the contamination. The New York State Senate also announced a series of hearings on "water quality" starting in August. The first session would be held not in the gilded sandstone chambers of the state capitol, but in tiny Hoosick Falls, where tensions were reaching a boiling point.[6]

MICHAEL HAD PLANNED TO SKIP THE HEARING. He already knew most of the ugly details that were about to be revealed to the general public, and the Gillibrand forum had left him even more skittish about being in the spotlight. But the morning before the proceedings were set to take place, organizers asked Michael and Marcus to give the opening testimony. Michael spent most of the day agonizing over what to say, while Angela—who worried his appearance would open him up to more attacks—paced about the house, feeling queasy.

When the hearings began, Michael and Marcus were seated on the stage of the school auditorium opposite two long tables packed with lawmakers. Michael was visibly sweating. When he launched into his meandering remarks, some onlookers strained to make out what he was saying.

But then he started talking about why he'd taken up this fight even though it went against his entire nature, and his voice grew stronger. It all went back to his father. The humble school bus driver had worked so hard his entire life, only to fall ill before he had a chance to enjoy his retirement, Michael said, and he began to weep. Michael added that he understood why people were angry at him for challenging the factories that many families relied on for their livelihoods. But the question driving him had always been, What would Ersel have wanted? "I think it goes back to the kids on the bus," he said, pausing to collect himself. "Because for him, that was his family as well. And I think that what he would have wanted was to protect them." Above all, Michael noted, he hoped the hearings would bring about a commonsense system for dealing with factory pollution so that ordinary citizens didn't have to "battle through the multiple levels of red tape" just to get clean drinking water.[7]

The hours of testimony from state and local officials that followed revealed little in the way of new information. Instead, the most illuminating statements came from residents who described how the crisis had affected them. It wasn't just the economic fallout or the health worries. All their lives, they had trusted that there were systems in place to protect them, and now that trust had been shattered. As Rob Allen put it, Hoosick Falls had "raised a generation of children . . . who don't trust the government because it largely failed them, and a generation of adults who have realized how to advocate and how necessary it is to do so."[8]

The hearing got Rob and Emily fired up. They wanted to take their activism to a new level, maybe even influence national regulation. But when they tried to enlist Michael, he told them he wasn't interested. That spring, a temporary filtration system had been brought online to remove PFOA from the public water supply.[9] Now that the water was safe and so many people were involved, Michael wanted his life back.

But the habits he'd developed over the last three years proved hard to shake. He was still up reading until the early hours, still attending all the public meetings, still fielding queries from sick residents. He'd also become fixated on proving that he'd been acting out of devotion to his family and community. That fall, when he and Angela got married, the wedding was largely a tribute to Ersel. Marcus and the other groomsmen wore orange golf tees on their lapels as boutonnieres. The officiating priest wove bits of Michael's testimony from the public hearing into the homily. ("That's who *we* are. We're about kids on the school bus. Always. Not selfishly, but selflessly looking out for each other.") After the ceremony, Michael whisked his bride to the cemetery in Ersel's Corvette and laid some orange golf tees on his father's grave.

Angela didn't protest. But in the months that followed, she and Michael bickered constantly about his preoccupation with the water. "I thought that after we got married, he was going to step back and we were going to be a family," Angela said. "It didn't work out that way."

THAT DECEMBER, Michael attended the regular monthly meeting of the village board. At first, the discussion focused on the usual humdrum topics. But then, about an hour in, Borge announced that the board was headed into a closed-door session to discuss an agreement he'd reached with Saint-Gobain and Honeywell, which had quietly joined the negotiations.[10] The trustees planned to vote on it that same evening.

Michael was floored. There'd been no mention of this deal in the agenda circulated before the meeting. When a local business owner asked to see a draft, Borge told him that it wasn't possible. Following a volley of angry questions from the audience, the board agreed to postpone the vote. But Michael sensed he had another fight on his hands. On his way out to the car, he called Dave Engel, who immediately tipped off the media and began hounding the village to release the agreement.

Eventually, the terms were made public: Saint-Gobain and Honeywell had agreed to pay the village $850,000, largely to cover its past legal and public relations fees.[11] In return for this pittance, the village would have to give up all future claims related to PFOA in its water system, regardless of how catastrophic the consequences turned out to be. Borge argued that this trade-off was worth it. The crisis had drained the village's coffers. Without a rapid infusion of cash, Hoosick Falls would have to take on major debt, saddling residents with higher taxes and water bills. But locals were furious that the village government was still maneuvering in secret and trying to sell them short.

As for Michael, he'd lost all hope that Borge would ever stand up to the polluters. With local elections just a few months away, he started plotting to overhaul the village government. Ever since seeing Rob's eloquent testimony at the August hearings, Michael had thought he would make a great mayor. Rob was already deeply invested in the water issues. And he was bright and charismatic in his own quirky way, with a knack for making people feel heard. So Michael pressed Rob to run.

Rob, a devoted evangelical Christian, was torn. Between work, family obligations, and his activism, he was already stretched thin. But after several weeks of prayer and contemplation, he came to the conclusion that being mayor was part of God's plan for him. By the time he declared his candidacy in early February, Mayor Borge had made it known that he wouldn't seek reelection, and there were no other contenders. So Rob threw himself into learning how to run a village. He met with the police chief and read up on municipal regulation. Speaking on his headset while washing the dishes or painting the house, he consulted lawyers and scientists about strategies for dealing with the village's water problems.

All the while, Borge and the village lawyers continued quietly negotiating with the companies. In mid-February, they released a new version of the agreement, which provided an additional $150,000 in compensation but still barred the village from filing future claims.[12] With that, the anger that had been smoldering throughout the winter erupted in white-hot rage.

On February 27, when the board gathered at the old armory to vote on the agreement, more than two hundred angry residents turned out, many wielding protest signs. ("Stop the Madness!" "Dirty Water, Dirty Deal!") Borge had to shove his way through a scrum of reporters to get to the board table. After the usual formalities, he read a plodding statement in support of the deal. The audience burst into jeers. "Why did you take longer to tell us the water was dirty than you're taking to make this decision!" one woman bellowed, as the trustees sank into their chairs. But Borge kept stolidly reading. His tenure as mayor would be over in a few weeks, and he was determined to pass the deal while he could.

When it came time for public comments, Rob approached the podium. As the future mayor, he was worried that the outpouring of fury toward village officials would cause them to stop listening to dissenters and simply vote the deal through. Just before the meeting, he'd put out a video urging residents to voice their opposition in a "neighborly" way.[13] Now he implored the trustees to consider *why* people were so angry. "This community feels backed into a corner, betrayed, ignored," he said. "This community needs desperately, more than anything, someone who is in control to stand up and say . . . 'I got your back.'" Regardless of the village's financial situation, locals had lost too much to accept a deal that was wildly out of proportion with the harm the companies had caused. "We've lost family members, we've lost months and years to treatments," he continued, fighting back tears. "So, I'm asking you guys to please have our back and vote this down."

Over the next two hours, dozens of residents begged the board to reject the deal. Among them was a seventeen-year-old girl who spoke on behalf of all the young people longing to "build this town up." "Please let us have our chance," she pleaded, "*Please* don't sign this agreement!"

Borge listened impassively to these appeals, then read a formal resolution to approve the deal and called a vote. Livid residents leapt to their feet screaming, *Shame on you! How dare you!* The police stationed around the room surveyed the crowd warily, as if they were bracing for a riot.

To everyone's surprise, it was Ric DiDonato who stopped things from getting ugly. As the mayor's deputy, he'd been among the deal's staunchest defenders during the months of public tumult. Now he grabbed the microphone and, struggling to be heard over the din, proposed that the agreement be tabled until the new village government was seated. Nodding toward Rob, he added, "I'm putting my faith in you, that you will be able to do a better job than I have." The room erupted in frenzied applause. The board voted to table the deal, with Borge the lone holdout, and the meeting finally adjourned. Then the defeated mayor stalked out of the building, guarded by four police officers.

SEVERAL WEEKS LATER, Rob Allen was elected mayor. He immediately set about unraveling the village's deal with Saint-Gobain. As he worked his way through Borge's numerous emails, the entire unsettling picture became clear. Borge had negotiated the agreement with the companies largely on his own. Even the other board members weren't privy to the talks. The lawyers who had ostensibly guided the process were brought on only after key details were worked out.[14] And most of the hundreds of thousands of dollars' worth of hours they'd billed had been spent trying to manage public opinion or thwart Healthy Hoosick Water. (In an email to Borge, one lawyer suggested that the quickest way to solve the village's problems would be to confront Dave Engel in a dark room with a "sock filled with nickels.")[15] As the public uproar grew, the lawyers also hired a PR firm that regularly worked with corporate polluters to prepare Borge's public remarks and draft the often-misleading materials the village put out to inform residents about water issues.[16]

Appalled, the trustees called an emergency meeting and voted unanimously to fire their law firm. They all agreed that whoever represented the village going forward should be well-versed in PFOA and willing to advocate aggressively for the town's interests. To one faction, the choice was obvious: Dave Engel. After some heated debate with trustees who'd been on the receiving end of Engel's sallies, the board voted overwhelmingly to hire him.[17] In a move that would have been unthinkable during the Borge era, they also gave Engel the authority to sue Honeywell and Saint-Gobain if negotiations stalled.[18] The village of Hoosick Falls was finally ready to fight.

15

Accidental Activists

One crisp evening in October 2016, Emily tore the Cloud Nine sign off the stump in her front yard and tossed it in a corner of the basement. She had tried hard to put on a brave face for her family, but the truth was that being in her home had become unbearable. With the smokestacks right next door, she knew toxic chemicals were permeating everything—not just the water but also the soil in her garden and the air her children breathed. Everything her family touched was poisoned.

Desperate to get them someplace safe, a few months earlier Emily had dug out the old FOR SALE BY OWNER sign that her children had retrieved from the driveway the day they'd closed on the house. She wrote her phone number on it and got Gwen to plant it back out by the road. She didn't get a single call. Eventually, one of her mother's realtor friends agreed to market the house as a favor, but he warned Emily that the water would make it a tough sell. Week after week, she scrubbed the house spotless for showings, but it didn't do any good.

Emily sank into a deep depression. She'd always taken such pride in keeping a clean and cozy home, even when her family had been crammed into the two-bedroom trailer. Now the hampers overflowed and the dishes piled up in the sink. She started sleeping during the day and snapping at her children. Emily was also plagued by peculiar ailments that seemed to come out of nowhere. Migraines. Debilitating stomach pain. Erratic periods with flow so heavy that one time the blood soaked right through her clothes, puddling on the floor. Her doctors eventually managed to pin down the cause of her symptoms: a combination of thyroid disease and endometrial hyperplasia, an abnormal thickening of the uterine lining that can morph into uterine cancer, a disease linked to PFAS.[1]

Emily was so distraught that she considered walking away from the house and filing for bankruptcy. "I don't have the energy anymore to play Russian roulette with my children," she explained. Michael did his best to help her. He was always ready to lend a sympathetic ear, lighten her mood with off-color humor, or brainstorm solutions to her accumulating problems. When Emily confided that her kids' pediatrician wasn't taking her concerns seriously, Michael persuaded Marcus, who wasn't technically accepting new patients, to take over the children's care. During their first appointment, Marcus outlined a detailed plan for monitoring her children's health. "Just sitting in the room with him, I felt so much relief," Emily recalled. That night, she was able to sleep without NyQuil for the first time in months.

Meanwhile, Emily's realtor called to say that a couple from Colorado was interested in her house. Emily tried not to get her hopes up. The few people who had inquired over the past year had all lost interest as soon as they found out about the water. But that same evening, the couple made an offer. It was less than what Emily had spent to buy and fix up the place, and they wanted her to throw in her furniture, too. ("Furniture wtf?" Michael joked in a text message. "Why not toss in a kid?")

Emily decided to take the hit and threw herself into searching for a new home. But even after weeks of scouring listings in her area, she couldn't find a decent place within her budget. When Michael suggested looking in Hoosick Falls, she was skeptical at first. Wouldn't it defeat the entire purpose of moving if they wound up in another polluted town? But then she got to thinking. After everything she'd been through, she didn't trust the water anywhere. At least in Hoosick Falls, she knew what kind of pollutant she was dealing with and there was now a permanent filtration system in place to remove it.

In early 2017, Emily toured a charming three-bedroom with golden oak floors on a peaceful cul-de-sac near town. It happened to have the same speckled brown countertops her grandfather had installed in her Petersburgh home: Just the sight of them gave her goosebumps. Believing it was meant to be, Emily put in a winning offer, then she and Jay began packing for the move. During the roughly four weeks between the sale of her old home and the closing on the new one, she arranged for each member of the family to stay with a different relative.

But then all her careful planning hit a major snag. The appraised value of the Hoosick Falls house dropped, probably because of the water problems, and the bank wouldn't approve a mortgage for the amount she needed. Emily panicked. What if they lost this house and couldn't find another one? Rather than take that risk, Emily agreed to pay the sellers the $12,000 difference out of her own pocket. Squeezing the extra monthly payment out of her already-tight budget would require scrimping on the children's Christmas presents and giving up household staples like hair conditioner, but at least her family would have a home.

On the day of the closing, Emily showed up to meet the sellers wearing cheap rhinestone sunglasses on her head—a vain attempt to tame her frizzy, unconditioned hair. Everyone signed the papers, then Emily was given the keys. Holding them tight, she waited for the relief to set in.

But it was only when Emily finally moved to Hoosick Falls in May that her anxiety ebbed. For the first time in months, she began to take pleasure in simple things, like stopping to chat with her new neighbors during meandering walks into town. Once she'd hung the family photos and stenciled her favorite adages on the walls, the new house felt even homier than the old one. Somewhere along the way, she began to nurture a tiny tendril of hope that her family's ordeal was finally coming to an end.

BY THIS TIME, PFOA had mostly been phased out in the United States as part of the voluntary deal between industry and the EPA. But the forever-chemical crisis was far from over. Most U.S. manufacturers that used PFOA had simply replaced it with lesser-known chemicals from the same family. Industry claimed these substitutes were safer because they passed more quickly through people's bodies due to their shorter carbon chains.* But the public had no way of knowing whether this was actually true. And while manufacturers had to supply the EPA with data on the chemicals' structures and properties, the companies could hide these details from the public—and even most EPA staff—by labeling them "confidential business information." (The identities of more than ten thousand chemicals are shielded under this provision, which is integral to the Toxic Substances Control Act.)[2] This meant that *another* generation of virtually indestructible chemicals was spreading through the environment before scientists even had the tools to detect them.

A few enterprising researchers went to elaborate lengths to unmask the mystery molecules. In 2012, Mark Strynar and Andy Lindstrom of the EPA's National Exposure Laboratory set out to reverse-engineer the formulas of chemicals polluting a stretch of the

* Both PFOA and PFOS contain eight-carbon chains swaddled in fluorine, while their replacements have seven or fewer consecutive carbon atoms.

Cape Fear River in North Carolina, near a plant where DuPont and its spin-off Chemours had been producing forever chemicals for decades. The scientists spent months trudging up and down the riverbank collecting samples. In the end, they concluded that the entire watershed was polluted with a stew of previously unidentified PFAS. Using a groundbreaking method that combined computer modeling with highly sensitive mass spectrometry, they managed to map the structures of the new substances.

Because Strynar and Lindstrom couldn't access the manufacturers' confidential formulas, they had no idea whether the molecules they'd discovered were new creations, manufacturing by-products, or older substances that had been kept secret. However, while combing through papers DuPont had filed with West Virginia regulators, they unearthed the formula for the company's main PFOA substitute—a substance known as GenX—and were able to match it with one of the chemicals they had isolated. "It was a sort of 'ah ha' moment," Strynar recalled. "Up to that point we were still grasping at straws."[3]

The researchers eventually partnered with scientists from North Carolina institutions to publish a paper with the formulas for a dozen previously unidentified PFAS, including GenX, as well as the methods used to identify them.[4] This enabled scientists around the world to do their own sleuthing. What they uncovered was a horror story.

It turned out the new generation of forever chemicals was already everywhere, from penguins in Tasmania to rain on the remote Tibetan plateau.[5] And these substances posed a far graver threat than industry had let on. A 2017 analysis by a group of Swedish scientists found that GenX was even *more* toxic than PFOA, a conclusion later confirmed by a formal EPA assessment.[6] Research also showed that the supposedly safer short-chain PFAS spread more rapidly through the environment and accumulated faster in crops, leading to higher concentrations in food.[7] It was also much harder to get these substances out of drinking water, since they didn't bind as readily to the type of carbon filters used to remove PFOA. The upshot: After a relatively short period, the filters disgorged the chemicals into the treated water supply.[8]

Despite the alarming implications, these findings barely registered outside scientific circles. Then, in early June 2017, a North Carolina paper broke the news that more than two hundred thousand people downstream of the Chemours plant were drinking water heavily polluted with GenX.[9] In the nearby city of Wilmington, the revelations sparked a panic. Hundreds of frightened residents crammed into public meetings or protested in the streets, catapulting forever chemicals back into the national headlines.

LATER THAT MONTH, Michael and Emily joined about a hundred locals in the Hoosick Falls Central School auditorium for an urgent meeting, called by Mayor Rob Allen. As part of New York State's response to the contamination crisis, Governor Cuomo had named PFOA a hazardous substance under its Superfund law. This was the first time a forever chemical had been hit with this designation, which empowered regulators to force those responsible to investigate the scale of the contamination and pay for cleanup. Now, it was time to unveil the preliminary findings. In his opening remarks, Allen likened the occasion to a doctor's appointment where you know you'll be getting bad news. "You just have to figure out *how* bad," he said glumly.

An official from the state Department of Environmental Conservation called up an aerial map of the village with six red blots along the river, marking the locations where Teflon had been used in manufacturing. Over the next two hours, he and representatives from Saint-Gobain and Honeywell unpacked the details: Investigators had collected hundreds of soil and water samples at varying distances from these sites. Virtually every specimen, including dirt from eighty feet deep, was saturated with PFOA. And the groundwater beneath the McCaffrey Street plant contained one of the highest levels of PFOA ever detected anywhere at the time.[10]

Sitting in the audience, Emily buried her face in her hands. It wasn't just the confirmation that her new hometown was awash in PFOA that worried her. She'd been following the news out of North Carolina and reading up on the replacement chemicals being detected in other factory towns. If her family was being exposed to them as well, they could face a whole new set of health risks.

After the presentation, Emily headed to the gym, where presenters had set up information booths, and marched right up to the Saint-Gobain station—a bare plastic table manned by the company's environmental project manager. Without any preamble, she asked him what chemical Saint-Gobain was using in place of PFOA. Seemingly flustered by this straight-talking mom in her old gray fleece, the official directed her to a dapper man in a purple pinstripe shirt—company spokesman Carmen Ferrigno. Ferrigno tried to send her back to the first man, but Emily wouldn't have it. "I just want to know what replaced PFOA," she insisted. "I want to be able to research it, because that's coming out of the stacks and landing now."

"So write down what you want to know," Ferrigno replied, his tone suddenly sharp. "You just asked me about what's replacing PFOA, and then you said something about what's coming out of the stacks. You're asking me a series of different questions." Emily calmly repeated her query. Ferrigno replied that Saint-Gobain wasn't always privy to its suppliers' formulas, then swiftly switched the subject, trumpeting the company's record of supporting environmental causes. "The climate change accords in Paris, we were the premier sponsor!" he said. "You see, we're the kind of company that cares about the communities we work in."

• • •

AFTER THE MEETING, two more Hoosick Falls factories and an abandoned landfill were named state Superfund sites, and the McCaffrey Street plant landed on the EPA's National Priority List of the worst hazardous waste sites in the country.[11] Realizing it could take years to fully investigate and address the contamination, Rob Allen and Dave Engel went on the offensive, demanding that Saint-Gobain cover its share of the village's contamination-related expenses on an ongoing basis without officials having to sign away future rights.

Once again, the company responded by emphasizing how crucial it was to the local economy. In August 2017, the CEO of Saint-Gobain's entire North American operation, Tom Kinisky, traveled to Hoosick Falls for a meeting with the mayor and his deputy at the McCaffrey Street plant. There, Kinisky's team unveiled a proposal to expand the company's Hoosick Falls operations, creating more tax revenue and jobs. [12] "It felt like a thinly veiled threat," the mayor's deputy recalled. "The implication being, 'Well, we could expand here or we could just leave.' "

Rob also pressed Saint-Gobain and the state officials overseeing the state Superfund investigation for information about the replacement chemicals the company was using in its local factories. When the state finally disclosed that the plants were using GenX, the mayor panicked. If the chemical was in the water supply, he warned trustees, the entire situation would "go nuclear." Lab tests that December found that GenX hadn't yet reached the public wells, but another toxic forever chemical had. Worse, it was already breaking through the first stage of the filtration system installed ten months earlier.[13]

Honeywell and Saint-Gobain agreed to switch the filters more regularly to keep the chemical out of the tap water. But there was no guarantee they would do so indefinitely; although New York State had made history by regulating PFOA, it had no authority over the thousands of other forever chemicals now in circulation. The only way to truly protect the village, Rob realized, was to heed Healthy Hoosick Water's advice and tap into an entirely new water supply, something that no PFAS-polluted community had ever managed to do.

Many of these towns were in far tougher situations. Researchers would later find 89 forever chemicals, including GenX, blowing from the smokestacks of Saint-Gobain's plant in Merrimack, New Hampshire. Another 250 were detected in Wilmington, North Carolina's raw water supply. Similarly, aquifers near U.S. military bases around the country were found to contain more than 200 different PFAS, many of them previously unknown to scientists.[14]

These mystery variants were also being detected in more-surprising places, including microscopic particles floating through the atmosphere and in the blood of Chinese newborns.[15] All of which raised serious questions about the value of the PFOA phase-out. As Lindstrom noted, "It's like we cut the head off the hydra, and it sprouted all these new heads."

IN MANY WAYS, the forever-chemical saga was following an entirely predictable pattern. Our regulatory system presumes most chemicals are safe and, when problems arise, investigates them one by one. As a result, when companies come under pressure to phase out a given chemical, they usually replace it with unvetted substances that have similar structures and properties. Not surprisingly, the substitutes often turn out to be just as harmful, though this can take years to establish.

In theory, the EPA was supposed to be fixing the system that created this toxic shell game. After many years of negotiation, in 2016, our famously gridlocked Congress voted overwhelmingly to reform the Toxic Substances Control Act, ushering in the most significant environmental legislation in decades. The revised law required the EPA to screen *all* new chemicals coming onto the market and—more importantly—mandated vetting for existing chemicals. No more grandfathering.[16]

The bill owed its success partly to backing from the chemical industry. Facing intense public anxiety over ever-present chemicals like PFAS and phthalates, manufacturers saw federal regulation as the best way to preserve faith in their products and to avoid a cumbersome patchwork of state regulation. In return for their support, industry lobbyists insisted on provisions that would make the process slow and ponderous. Still, environmental groups hailed the law as a major victory that they could build on over time.[17]

What they hadn't reckoned on was a Donald Trump presidency. Almost as soon as the new administration took office, it began working to gut the legislation entirely. Just weeks before the EPA was due to finalize its rules for implementing the law in June 2017, a former American Chemistry Council lobbyist named Nancy Beck was appointed to the number two spot in the agency's Office of Chemical Safety and Pollution Prevention. Beck pressed agency staff to rewrite the rules along industry-friendly lines. One of her chief demands involved guidelines for picking which chemicals to prioritize for safety testing and possible regulation. The draft called for factoring in various ways people are exposed, since heavier, more widespread exposure can translate into a greater risk. Beck wanted to overhaul the language for "legacy" chemicals like PFOA that had theoretically been phased out. Specifically, she wanted to bar the EPA from considering key routes of exposure—including drinking water.[18]

Beck's colleagues warned that this approach would drastically underestimate the hazards since these chemicals were still ubiquitous in the environment despite dwindling U.S. production. Beck ignored their concerns and took charge of rewriting the rules herself, a highly unusual move at an agency where guidelines are usually drafted by career civil servants with specialized expertise.[19] President Trump, meanwhile, tapped a former Saint-Gobain lobbyist to head the EPA office that oversees the federal Superfund, one of the programs Hoosick Falls was counting on to compel polluters to pay for cleanup.[20]

Michael followed these developments compulsively, fearing they would erode his community's hard-won gains. The question was what to do. He knew his usual approach of maneuvering behind the scenes wasn't going to cut it now that he was up against the opaque workings of the federal bureaucracy. So when the regional Democratic Party approached him about running for the county legislature, he was tempted. The county played a crucial role in responding to environmental crises, including the one in Hoosick Falls. If he managed to get elected, he could shape the response going forward and tap into official information, which might give him leverage on the national level. Angela warned him that he didn't stand a chance in their deep-red county. But Michael was hopeful that his dogged advocacy would count for something with conservative voters.

Politicking didn't come naturally to Michael. He could be awkward at campaign events, and he would turn up to canvass working-class areas in his BMW and preppy clothes, invariably rubbing some voters the wrong way. Michael himself found the process torturous, but Trump kept providing him with fresh motivation. In July, the president announced his nominee to head the EPA's Office of Chemical Safety. It was none other than Michael Dourson, the industry-funded toxicologist who helped set the dangerously lax standard for PFOA in West Virginia. If confirmed, Dourson would not only oversee implementation of the new chemical-safety law but would also be in charge of vetting some of the very substances whose dangers his group had downplayed while taking money from their manufacturers.[21]

Dourson's selection triggered a ferocious backlash, and not just from environmental groups. Labor leaders—incensed by Dourson's role in diluting workplace safety rules—joined forces with health advocates, research scientists, farm workers, and medical organizations to try to sink his nomination.[22]

Their best chance was to influence the vote in the Senate Environment and Public Works Committee, the hurdle Dourson needed to clear before the full Senate vote. Several controversial Trump picks had withdrawn under pressure *before* the committee vote on their nomination. But every nominee who won committee approval had been confirmed by the full Senate. Dourson's opponents would be up against the entire U.S. chemical industry and its armies of lobbyists, but they had powerful allies of their own. In the weeks before the pre-vote hearing, environmental groups reached out to dozens of ordinary Americans who had been harmed by the chemicals that Dourson's group defended, and invited them to Washington to tell their stories.

THE MORNING BEFORE THE HEARINGS, Michael arrived at the Hart Senate Office building wearing a navy suit, wingtips, and orange lightning-bolt socks. He was followed by Loreen Hackett, who carried photos of her two small grandchildren holding signs with their PFOA levels, then Emily and her twelve-year-old daughter, Gwen. Brimming with excitement, the group was ushered into a cramped conference room, where, for the first time, they met people from all over the country who understood exactly what they were up against. There was a young man whose childhood battle with brain cancer inspired a federal program to track pediatric cancer clusters, and the gray-haired engineer who had managed Little Hocking's water district and spent fifteen years fighting to get DuPont to clean up its well field.

When it came time for introductions, a burly former marine sergeant whose family had drunk tainted water while he was stationed at Camp Lejeune, described losing his nine-year-old daughter to cancer. A young woman from Wilmington, North Carolina, held up a photo of her dead newborn wrapped in his baptismal gown. His kidneys and bowels simply hadn't developed, a "symptom," she said, of the GenX she'd absorbed during her pregnancy. A gaunt fifty-seven-year-old Parkersburg woman named Sandra described how the explosive diarrhea from her ulcerative colitis had made her afraid of going out in public.

After the formalities, Emily made a beeline for Sandra, who had been fasting in preparation for the trip and was wistfully eyeing some cinnamon-crumble muffins on a nearby table. While the women chatted about the "tummy tricks" Emily's mother used to ease her own ulcerative colitis, Gwen circled the room snapping selfies and hugging strangers. Michael retreated to a corner and sat kneading his sweaty hands.

Around nine, the environmental groups split the activists up by state, assigned them guides, and dispatched them to their first meetings. The Hoosick Falls contingent was paired with a legislative attorney named Melanie Benesh, who passed out cough drops to help with all the talking they'd be doing. Some lawmakers would be sympathetic to their cause, Benesh explained; others would need more persuading. Ideally, she said, they would manage to convince some of the committee's conservative senators to delay the vote on the grounds that Dourson was politically "toxic."

For the rest of the morning, the group crisscrossed Capitol Hill, meeting with mostly Republican staffers. Not all of them were receptive. An aide to Senator John McCain of Arizona opted to take the meeting in a packed hallway. Hands trembling, Michael launched into the story of his father's death, but he was drowned out by the din of voices and hard-soled shoes echoing through the corridor. The aide just stared at him blankly; Michael walked away, deflated.

Yet a few minutes later, he was sitting in the chambers of Senator Susan Collins, the Maine Republican, for a meeting with legislative director Olivia Kurtz. Feeling more at ease, Michael recalled how he and Marcus had struggled for years to get the village clean drinking water but had only made real headway when the EPA finally intervened. With people like Dourson at the helm, Michael argued, the agency wouldn't do that. Small towns in rural areas like upstate New York or Maine would have to fend for themselves against giant corporations. Kurtz nodded sympathetically, scribbling notes.

Then Loreen pulled out the photos of her grandkids with their PFOA levels and detailed their various health problems: autism, seizures, bone disease, migraines, weakened immunity. Emily told the story of her polluted well and how she'd been forced to sell her home and uproot her children. "The American dream was ripped out from under us," she said. Nodding toward Gwen, she talked about the lumps in her twelve-year-old daughter's breasts and her own decision not to have another child. "This chemical isn't leaving my body and it would be passed on, which is scary," Emily said, pressing her hand to her heart. Kurtz, who seemed genuinely moved, thanked them for coming and promised to share their concerns with Senator Collins.

By the time the Hoosick Falls crew left the capitol complex that evening, they had logged more than six miles and everyone looked spent. Loreen's makeup was smudged, and Gwen was hobbling, her feet blistered from her new ballet flats. Even the usually dapper Michael looked a bit rumpled, with the tail of his gingham shirt untucked and flapping in the wind.

THE FOLLOWING MORNING, lobbyists and reporters jostled for prime positions in the committee hearing room. Picking their way through the crowd, Michael, Emily, Gwen, and Loreen joined the other community activists packed into the back few rows.

When the hearing got underway, the committee's ranking Democrat, Senator Tom Carper, warned that Dourson's confirmation would make a mockery of the rare bipartisan victory that the committee had celebrated just one year earlier with the passage of the chemical-safety bill.[23] Standing behind Carper, an aide hoisted a sign with the headline "Science for Sale." It listed safety limits that regulatory agencies had set for various chemicals, alongside the much higher values Dourson's group had picked for the same substances. Democrats took turns grilling Dourson, who sat bolt upright at the witness table, seemingly rattled by the questioning and the swarm of cameras.

When it was New York Senator Kirsten Gillibrand's turn, she acknowledged Michael, Emily, Loreen, and Gwen by name. "These families are so frightened," she said, choking up. "I can't imagine what it would be like to live and not know if the water that your children are being bathed in is safe; if they are going to get *cancer* when they are twenty-five; if they are ever going to be able to have kids." Addressing Dourson, she added, "Their lives are so affected by the decisions that you have made, and I don't think you recognize, when you are hired by a company, when you are hired by the DuPonts of the world . . . you are being asked to change how governments—how leaders—look at these risks and whether they say it is safe or not."[24] Wiping away tears, Emily blew Gillibrand kisses.

When the committee met again later that month, the GOP majority voted unanimously to advance Dourson's nomination. Michael and Emily were heartsick: They assumed the Senate would simply rubber-stamp the committee decision, as it had done with every other Trump nominee.

But the surge of citizen activism was changing the political calculus in ways almost no one expected. That fall, Senator Joe Manchin of West Virginia, a conservative Democrat who had supported most of Trump's nominees, announced he would vote against Dourson, citing his state's own troubles with PFOA. North Carolina Republicans Richard Burr and Thom Tillis followed suit, pointing to the pollution from Chemours's plant near Wilmington. Susan Collins disclosed that she was "leaning against" Dourson, too.[25]

Suddenly, Dourson's nomination was in serious trouble. Hoping to reassure skittish conservatives, the Trump EPA unveiled a detailed plan to tackle the PFAS crisis. This was a bold move for an administration that was otherwise bent on demolishing regulations. But it wasn't enough. With defeat looming, that December Dourson withdrew his nomination, handing the growing brigade of grassroots activists their first national victory.[26]

16

What-Ifs and Worst-Case Scenarios

As public awareness of forever chemicals grew, Michael found himself flooded with inquiries from reporters and advocates across the United States and around the world, as other countries started waking up to their own PFAS crises. He was featured in news stories from Australia to Holland and delivered a virtual lecture to a college in Israel.

But back home in Hoosick Falls, many people still suspected that he was only in it for the money. As the county legislative election drew near, his opponents played on this distrust, sending canvassers door to door to remind voters about his class-action suit against Saint-Gobain. Michael hoped his community would see through this. But on election night, when he arrived at a local bar to watch the returns, he noticed that many people wouldn't look him in the eye.

His mood only deteriorated as the results trickled in and one municipality after another was called for his opponents. When the numbers for Hoosick Falls arrived and Michael realized he'd lost there, too, he could barely believe it. How could locals see him as so unworthy after he'd worked so hard to protect them? Devastated, he started pounding shots and loudly railing against the winning candidates.

In the weeks and months after the election, Michael couldn't get over his hurt and anger. He started drinking more heavily, sometimes mixing booze with stimulants to keep his benders going longer. Often, Angela would stay awake until 3 or 4 a.m., too nervous to sleep, waiting for him to call her from the bar for a ride and worrying what would happen if he didn't. "Even if he just gets pulled over for a DWI, they're going to be running his mugshot on the five o'clock news and talking about the water hero driving under the influence," she said. "It could sabotage everything he's been working for." But when she tried to confront Michael about his drinking, he got defensive. Sometimes, they'd wind up in a vicious argument that ended with Angela slumped at the foot of the bed sobbing and Michael stomping out to the bar.

Desperate, Angela started turning up at the bar to try to drag her husband home in the wee hours of the morning. On weekends, she made plans to keep Oliver out of the house, so he wouldn't see his father surly and hungover. Simmering away beneath this strife was their mutual resentment over the water, which Angela felt was consuming their lives and sabotaging their marriage. For his part, Michael was bitter that his wife hadn't been more supportive, especially after his crushing election loss. Marcus, who was growing increasingly concerned, pleaded with his friend to seek counseling. Michael was too ashamed to do so, but he knew that something in his life had to change.

EMILY WAS STRUGGLING, TOO. The previous December, she'd started vomiting up her morning coffee. After several days of this, she snuck into the bathroom with a pregnancy test and watched with horror as a plus sign appeared behind the plastic window.

A few years earlier, Emily would have been thrilled to bring another child into the family. Now, what-ifs and worst-case scenarios hit her from every direction as she agonized over whether to have the baby. She felt she could deal with most of the problems that might arise; with Marcus's help, she could ensure that any diseases were caught early. But ever since she read about Sue and Bucky Bailey, she'd been haunted by the details of all his painful surgeries. Emily knew from scouring DuPont documents that the level of PFOA in her blood was even higher than Sue Bailey's had been when she gave birth.[1] And she wasn't sure her family could handle the heartache or expense of raising a child who needed such intensive medical care. "I mean, I'm already supporting two other children on a limited budget," Emily said. "If I had a child that required surgery after surgery after surgery, how would I do it? How *could* I do it?"

In the end, Emily decided to go ahead with the unplanned pregnancy—she didn't feel she could live with herself otherwise. But she was racked with anxiety. During one early appointment, she grilled her obstetrician about a special test to detect facial birth defects. The doctor suggested starting with a regular ultrasound. Emily nodded warily, and the doctor pressed the Doppler into her belly, amplifying the baby's heartbeat. Suddenly, Emily panicked and leapt up from the table, tugging her sweatshirt down over her gel-slathered stomach. "Oh, you're goopy!" the doctor said, handing her a tissue. But Emily didn't seem to hear her.

After the appointment, Emily sped home through the rugged Vermont hills. "I just hope I bond with this baby," she muttered aloud as she drove. This thought had never even crossed her mind during her earlier pregnancies, but she had fostered bonds with those babies through breastfeeding. And she had serious doubts about nursing this one, since she knew she would be passing PFOA on through her milk. When she consulted Marcus, he agreed that the risks likely outweighed the many benefits, given her blood levels. Having a clear, informed answer helped put Emily's mind at ease. So did her five-month ultrasound, which showed that the baby's heart and organs were healthy, and that she was having a girl.

But her nerves rushed back during the run-up to the final scan, which would show the baby's features in greater detail. When Emily arrived at the doctor's office with Gwen and Jay, her brow was furrowed and her face was broken out. She was ushered onto a table in a dim exam room that smelled of vomit and disinfectant. A technician moved the wand over her belly, measuring the baby's limbs and organs. Emily watched the grainy shapes appear and dissolve on the screen. As they neared the head, she begged to see her child's face, but the baby squirmed away from the Doppler, a blur of knees and vertebrae. The technician tried again, digging the wand into Emily's lower abdomen, but this time the baby's face was hidden behind a shoulder.

Finally, the baby's cheek appeared, followed by her nose and her tiny bow mouth. "Oh my goodness, look at those lips!" Emily cried. "They look perfect! And her nostrils! And her eyes! Perfect, am I right?" Gwen clapped her hands and squealed in delight.

WHILE EMILY WAS SHUTTLING BACK and forth to doctors' appointments, tens of millions of Americans were learning their drinking water was polluted with PFAS. In some cases, the information came from government agencies like the Pentagon, which had begun testing for the chemicals around military bases nationwide.

But in large swaths of the country, it had taken citizen sleuths like Michael Hickey to bring the contamination to light. Perhaps the most dramatic example involved a group of residents in Rockford, Michigan. Led by a local piano teacher, the crew had spent years investigating a former tannery that was shut down with minimal cleanup, even though tanneries are known to use a slew of toxic chemicals. They eventually tracked down a whistleblower inside the state environmental agency, who disclosed that the site was likely saturated with PFAS from Scotchgard, which the tannery had used by the ton. One of them also interviewed a former waste hauler, who recalled dumping truckloads of tannery sludge at various sites, including a landfill in a leafy neighborhood on the outskirts of Rockford, where the homes relied on private wells.[2]

Finally, in early 2017, the group alerted state environmental officials, who ordered the former tannery's owner to analyze the wells. Sure enough, many of them were massively contaminated. One of the samples registered 61,000 parts per trillion of PFOA and PFOS combined—by far the highest level ever detected in drinking water.[*3] Horrified, residents lobbied lawmakers and made sure the story was broadcast far and wide.

Eager to avoid a repeat of the Flint crisis, Republican Governor Rick Snyder responded by forming a PFAS Response Action Team to tackle potential forever-chemical pollution across the entire state. By the spring of 2018, the state was systematically analyzing all public water systems.[4] What it eventually uncovered was the stuff of public health nightmares. At least one-fifth of Michigan residents relied on drinking water that was sullied with PFAS. And some of the levels were much higher than the EPA's safety guidelines, which scientists considered far too lenient. Reporting on these findings, the *Detroit Free Press* described forever-chemical pollution as the "most widespread, serious environmental crisis" to hit Michigan since the 1970s.[5]

Of course, it wasn't just Michigan. As other states began systematically testing their public water supplies, the results were strikingly similar. Roughly two-thirds of all Californians relied on PFAS-tainted drinking water. So did about four-fifths of New Jersey residents. Worse, preliminary studies suggested that these substances were polluting nearly half of all water systems nationwide.[6]

---

* By contrast, Hoosick Falls had about 600 parts per trillion in its raw water supply.

. . .

Emily's daughter, Eliana Lynn Young, was born on July 28, 2018, just over six pounds with a full head of hair. In the recovery room, Gwen cradled her sister in her arms and fanned her face to keep herself from crying. "Oh, she's so tiny!" Emily's mother enthused when her turn came to hold the baby. Emily shot her a worried look: It troubled her that Ellie was several pounds lighter than her other children had been as newborns.

But then the nurse came in and laid the baby in Emily's arms, and her fears seemed to evaporate. Holding the swaddled infant close, she caressed her plump pink cheeks, admiring her delicate features and her pale blue eyes. "She's got the cutest lips," she murmured, and a wave of joy washed over her.

After the birth, Emily decided to take a break from activism to enjoy those precious early days with Ellie. But a mere two months later, she found herself rushing to the emergency room in Bennington with her nineteen-year-old stepdaughter, Sahara, whose blood pressure had suddenly spiked. The emergency room doctors determined that Sahara's kidneys were failing, and dispatched her by ambulance to a hospital in Albany. There, Sahara was diagnosed with chronic kidney disease and informed that she'd already lost most of her renal function and sustained some heart damage.

Emily couldn't help but wonder if PFOA was to blame. Sahara had spent less time in Petersburgh than the other children and thus had less exposure, but there was no other obvious explanation for her kidneys deteriorating at such a young age. For now, her symptoms could be managed with drugs and dialysis, but it was just a matter of time before she needed a kidney transplant, which raised more excruciating questions. Some of Sahara's closest relatives were at high risk of kidney problems because of their own PFOA exposure. Donating a kidney to her now could cost them their lives down the line. Emily found herself agonizing over what they should do if Jay, who had the highest levels, was also the best match.

Almost as soon as Sahara was back home, Ellie came down with a high fever and a barking cough. She wound up being hospitalized with a severe respiratory infection, stoking Emily's fear that her immune system was damaged from PFOA exposure in the womb. Gwen developed stabbing stomach pains, along with bouts of depression and fatigue. Blood tests turned up a marker of thyroid disease, one of the linked conditions that Emily suffered from.

In August, Emily took her entire family for the second round of state-funded blood testing. She hated pinning her squalling baby down so the nurse could poke the needle into her chubby arm, but she was desperate to find out Ellie's blood levels. When the stack of letters from the health department arrived that November, Emily heaped them on her kitchen counter and ripped open the one with Ellie's name. On her first read, she could hardly take it in. Her tiny daughter, who had entered this world just five months earlier, had 75,000 parts per trillion PFOA in her blood—far higher than most people in Hoosick Falls, where the average was about 28,000 parts per trillion.[7] This meant she would be vulnerable to serious health problems for the rest of her life.

The other envelopes contained more grim news. Everyone's PFOA levels had dropped by roughly half, as Emily expected, given that they hadn't been drinking polluted water for the last two years. But Jay's and Gwen's PFOS levels had spiked, and Gwen now carried a close cousin of PFOA that hadn't been in her blood before. Emily found it maddening that she had no idea where these chemicals came from.

Once she'd digested the numbers, Emily spread the printouts on the counter and picked up Ellie's again. "How is hers so fucking high?" she said, wiping away tears.

. . .

EMILY NOW FOUND HERSELF WORRYING even more about Jay, who had the highest levels of anyone in the family but hadn't yet been screened for linked diseases. So in January 2019, she asked Marcus to take him on as a patient. Marcus agreed, but he urged her to make an appointment soon, while he was still able to practice medicine.

This warning wasn't based on anything tangible. Sometime back, cancerous tumors had sprouted in his adrenal glands, but he'd managed to keep them in check, and they had no effect on his daily life. He still carried a full patient load and went on regular golfing excursions with his buddies. After his divorce a few years earlier, he'd started dating a spirited blond nurse named Gretchen, who had two daughters, aged four and seven. Although Marcus was taken with the girls' sweet, spunky personalities, he'd initially tried to keep his distance, hoping to shield them from grief when the cancer inevitably spread. Gradually, though, his defenses melted away. Gretchen and the girls had been spending weekends at his house, splashing in the pool, singing karaoke, and cuddling on the sofa watching reality TV. The couple were even talking about getting married and having a baby, something Marcus had always longed to do.

But that winter, Marcus had begun to sense that something wasn't right. Sure enough, one morning in February 2019, he collapsed on his kitchen floor. He came to and staggered into the living room, where he began vomiting uncontrollably. A scan the following day turned up two large brain tumors. As it turned out, the cancer had morphed into small-cell carcinoma, which was particularly deadly—most patients lived only a matter of months after it moved to the brain. However, the disease sometimes responded to novel immunotherapy drugs. Marcus decided that combining this approach with other aggressive treatments offered him the best hope of survival. But he knew his chances were slim. Bracing for the worst, he made his arrangements at the funeral home.

At the same time, he and Gretchen continued building a life together. She and the girls moved in with Marcus. They adopted a goldendoodle puppy and redecorated the living room, covering the walls with family photos and the hopeful sayings Gretchen lived by. ("HEAL the past. LIVE the present. DREAM the future.") When Marcus began the first phase of treatment, the physical effects were mild, but it sapped him of motivation. Often, he spent hours just staring into space, and he struggled to follow through even on simple plans. But there was one plan he was determined to carry out. One May evening, Gretchen arrived home to find the girls zipping around the house in excitement. She found out why when Marcus got down on one knee and slipped a ring on her finger.

The next phase of treatment, immunotherapy and chemo, hit Marcus much harder. He was so depleted by constant vomiting that he could barely climb the stairs to his bedroom. Then his white blood cell count plunged, and he started having trouble breathing. He was in and out of intensive care with blood clots in his lung and leg, a potentially deadly situation for cancer patients. By his final stay, he was curled up in a fetal ball and could only muster the strength to open one eye. His family was sure he was going to die. But Marcus survived—and the next scan showed that his brain tumors had shrunk. Yet again, he'd managed to beat the odds.

17

Wall of Resistance

By the time Marcus finished treatment, the groundswell of grassroots activism that he had helped set in motion was translating into actual political change. As part of New York State's ongoing response to the Hoosick Falls crisis, in 2018 it proposed binding caps of 10 parts per trillion each for PFOA and PFOS in drinking water, the toughest standards being considered anywhere.[1] California, Minnesota, New Hampshire, Michigan, New Jersey, Vermont, and Rhode Island were working on their own strict limits. And bills banning PFAS in food packaging and consumer goods were cropping up in statehouses nationwide.[2]

There was also progress in gridlocked Washington. In early 2019, nearly sixty members of Congress formed a bipartisan task force on PFAS. The group put forward an audacious bill that would obligate the EPA to designate PFOA and PFOS hazardous substances, giving the agency power to investigate polluted sites across the country and require those responsible to fund elaborate cleanup programs. The lawmakers also pressed for immediate caps on the two chemicals in drinking water.[3]

That February, Michael returned to Washington—Congressman Antonio Delgado, a New York Democrat, had invited him to be his guest at Trump's State of the Union address. Michael was struck by how easily doors opened to him now that forever chemicals had gained political traction. He and two other advocates were invited to meet with top EPA officials. He also met privately with senior Republican staffers from the Senate Environment and Public Works Committee, which oversees the EPA. After his previous lobbying experience, he wasn't counting on a warm reception. But the men paid close attention as Michael spoke about the people he'd lost and the need for commonsense solutions, like requiring companies to disclose their PFAS emissions so authorities could monitor for them.

Over the next few months, both chambers of Congress introduced a bumper crop of legislation. There were measures requiring safety testing and emissions reporting for all existing PFAS as well as one barring manufacturers from bringing new types to market. There was a provision requiring the EPA to set binding caps for PFOA and PFOS in drinking water and another banning fluorinated firefighting foam on military bases.[4]

Once again, environmental groups invited residents of polluted communities around the country to Washington to lobby for these measures. This time, Michael declined. He had recently separated from Angela. At age forty, he was living in his mother's attic and feeling utterly defeated, so he wasn't exactly eager to be in the public eye. But Emily and Gwen accepted the invitation, and Emily agreed to testify on behalf of residents of contaminated communities across the country at a congressional hearing in mid-May.

The night before the hearing, Emily and Gwen attended a reception for community advocates at the U.S. Capitol. It had been a year and a half since they'd first come to Washington to fight Dourson's nomination, and as they crested the marble stairway leading to the auditorium, it became apparent just how much had changed. More than a hundred people milled about the chamber—lawyers, politicians, professional activists.

Emily craned her neck, looking for Sue and Bucky Bailey. Ever since she'd gotten pregnant with Ellie, she'd thought about them constantly. Eventually, Emily spotted them and went to introduce herself. Holding hands, she and Sue traded stories about their family's struggles. At one point, Sue peeled back the corner of her wig, revealing her stubbly scalp, and detailed the ailments she blamed on DuPont, from brittle bones to Parkinson's disease.

Bucky, who was now in his thirties, had more hopeful news. Pulling out his phone, he showed Emily photos of his two young children: a slight, towheaded toddler and a chubby baby in a pink romper. Before they were born, he had worried that the effects of his exposure might be passed on. But, he said, the kids were healthy and so was he.

As usual, Emily was searching for clues about what awaited her own family. She asked Bucky how long after birth he'd drunk the polluted Parkersburg water and what his blood levels were. But Bucky had only been tested once as a baby, and neither he nor Sue remembered the number. For a moment, Emily looked crestfallen—even the Baileys didn't have the answers she so urgently needed. But then Sue slipped Emily her phone number and told her to call anytime, and her mood rebounded.

The following morning, Emily arrived at the House committee chambers wearing her old fleece and pink tennis shoes, and settled behind the witness table. Face flushed, she held up a picture of Ellie and told lawmakers that her cherub-cheeked baby had PFOA levels nearly forty times the national average, a burden she would not have borne if the government had acted sooner. "Congress needs to treat this as a crisis because it *is* a crisis," she pleaded. "I mean, I couldn't breastfeed. I couldn't do the most basic thing a mother does for my child because I knew it would elevate Ellie's levels."[5]

As Emily was speaking, more than two dozen advocates from thirteen states fanned out across Capitol Hill. Among them were a Maine dairy farmer whose century-old family business had been ruined by PFAS-soaked sewage sludge spread over his fields as fertilizer; a Colorado man who had seen five family members diagnosed with kidney cancer; and a young mother whose breastfed toddler had some of the highest PFAS levels ever detected in a child's body.

Amid the flurry of congressional hearings that followed, both chambers of Congress attached a raft of PFAS provisions to their versions of must-pass defense-spending legislation. These measures enjoyed remarkable bipartisan support: Nearly seventy members of Congress from both parties warned in an October letter to colleagues that they would not vote for *any* version of the bill that didn't preserve tough PFAS provisions.[6] After nearly two decades of inaction, it seemed national regulation was finally at hand.

BEHIND THE SCENES, THOUGH, these reforms were running into a wall of resistance from industry, which would spend tens of millions of dollars lobbying against them. The biggest corporate player, 3M, also dumped huge sums into congressional races and political action committees.[7] In addition, it cofounded a group called the Responsible Science Policy Coalition, which presented itself as an independent organization, though its briefing materials drew on the same skewed studies that 3M produced during its effort to "command the science" and create "defensive barriers to litigation."[8] The group dispatched lobbyists armed with these papers to persuade Congress and the EPA that there was no meaningful evidence that forever chemicals were harmful.[9]

Because these substances were so integral to American industry and military operations, many other powerful interests fought to kill the legislation—among them the Pentagon and the fossil fuel industry, which relied on PFAS for fracking fluid.[10] The hazardous-substance provision also drew fierce opposition from lesser-known organizations, including the American Water Works Association, whose members had unknowingly spread PFAS through wastewater discharge and sewage sludge, and faced potentially massive liabilities as a result.[11]

As pressure mounted, the tone in Congress began to shift. That September, the House Committee on Oversight and Reform convened another hearing, supposedly to investigate the chemical industry's decades-long PFAS cover-up. Rob Bilott, who had represented the West Virginia farmers, was called to testify, along with representatives from DuPont, 3M, and Chemours.

In a scene reminiscent of the infamous tobacco hearings of the 1990s, Democratic lawmakers grilled the corporate executives about their deceptive methods. An elaborate game of deflection followed. DuPont's chief operations officer, Daryl Roberts, claimed he was completely unaware of this history; it belonged to the *old* DuPont, which had ceased to exist after the 2017 Dow merger. The president of Chemours's fluorochemical division said his firm knew nothing either, since it was just an entity DuPont had created to "dump its liabilities."* 3M's representative, meanwhile, insisted her company had *always* been transparent, even as she made misleading claims about the health effects of PFAS.[12]

The spectacle infuriated Democrats—the committee's chair, Harley Rouda, called 3M's testimony a "slap in the face."[13] But despite the bipartisan concern on display in previous sessions, Republicans didn't challenge the executives' misleading claims. Instead, some lawmakers used their time to tout the countless technologies and lifesaving medical devices made possible by forever chemicals. Others suggested that Bilott's two-decade quest to call attention to

* Earlier in the year, Chemours had sued DuPont, claiming it had deliberately underestimated the PFAS liabilities it passed on to Chemours as part of its effort to evade responsibility for its legacy of pollution.

the problem was driven by financial self-interest. The ranking Republican, James Comer, pointed out that Bilott's firm earned about $22 million from the West Virginia class-action settlement, and demanded to know how much more Bilott made from the surge of PFAS cases that followed. "I'm not here to talk about me," the lawyer rejoined. "I'd like to talk about what we learned from these companies and why there's now a public health threat."[14]

That same month, the House and Senate began working to reconcile their versions of the defense-spending bill. When it came to PFAS provisions, the differences were mostly minor. By December, lawmakers were nearing a compromise that would have encompassed almost all the PFAS measures, including the controversial hazardous-substances designation. But as the deadline for finalizing the bill neared, Republicans began pushing back on the stricter provisions, arguing they would place an unfair burden on the military and private companies. At the last minute, these measures were stripped out of the bill entirely.[15]

By then, however, a global crisis was brewing that would eclipse almost every issue on the Washington agenda, PFAS included. In January 2020, a mysterious new virus began spreading through China, killing hundreds of people and throwing entire cities into chaos. With Covid-19 spilling across borders, later that month the World Health Organization declared a global "health emergency."[16] Executives at DuPont quickly realized that the world was about to face a shortage of protective gear for health-care workers. So they assembled a crisis team to speed up production of DuPont's protective suits, including coveralls made from its patented PFAS-coated Tyvek.[17]

The main hurdle was transportation: It took several months to ship the Tyvek to and from Vietnam, where it was sewn into suits. But in keeping with its long tradition, the company found a solution in the U.S. government. That April, as the virus was tearing through the United States, the White House agreed to fund chartered flights that could deliver nearly half a million suits every week, part of a pandemic supply effort called Project Airbridge.[18] The program, which was dogged by logistical problems, did little to ease the dangerous shortages of protective garments in hospitals around the country. But it was a huge success for DuPont. Thanks to the sponsored flights, the company was able to dramatically ramp up production of its coveralls, millions of which were sold to the federal government at inflated prices.[19]

18

Victory

One balmy afternoon in April 2021, Michael was on the bleachers watching Oliver's Little League game. From his perch on the top row, he could see a wisp of smoke drifting from the McCaffrey Street stacks and the nubs of the village wellheads poking up from the field below. Recently, Michael had begun to find some equilibrium. After more than a year of precarious living situations, he'd moved into a comfortable apartment with his girlfriend, Megan. And he'd been spending more quality time with Oliver, taking him to baseball games and theme parks up and down the East Coast. Sometimes, Michael got frustrated that after six years, he was still dealing with the water issues—but he was making some progress on that front, too.

For starters, he'd scored a crucial victory in his 2016 lawsuit seeking medical monitoring and compensation for Hoosick Falls residents. Around the time Michael filed, other locals had brought similar suits, and these had been merged into a single class action against Saint-Gobain, Honeywell, DuPont, and 3M. In early 2020, a federal court had ruled that the entire case could move forward, including the medical monitoring portion—even though the law required people to prove they'd been injured before seeking damages. The judge had embraced the plaintiffs' argument that simply being exposed to a chemical like PFOA was a form of injury, since it accumulates in the body and increases the risk of illness.[1] This potentially paved the way for similar lawsuits in other states that barred "no-injury" claims.

There were other promising developments on the legal front, too. The state supreme court had greenlit Emily's case seeking medical monitoring for the people of Petersburgh. A class action brought by North Bennington residents against Saint-Gobain was moving toward trial.[2] The village of Hoosick Falls had reached a series of interim agreements with Honeywell and Saint-Gobain to cover its pollution-related expenses. In a stark reminder of how far the negotiations had come, the companies had already paid out more than $1 million without the village having to sign away any future rights. And they were working with Mayor Rob Allen and attorney Dave Engel on a final settlement to cover the village's costs in perpetuity.

As Michael waited for Oliver to take his turn at the plate, his phone buzzed: It was his lawyer calling with some major news. Attorneys for the two sides in the Hoosick Falls class action had begun settlement talks. DuPont had walked out on the negotiations, but Saint-Gobain, Honeywell, and 3M were ready to discuss terms. It could still take months of haggling to reach an agreement, the lawyer explained, but a deal was actually within reach.

Michael's gaze fell on Oliver, now a winsome, dark-eyed eleven-year-old, sitting in the dugout. It felt like a lifetime ago that Michael had taken his birth certificate to Ersel's hospital room. The family had seen so many setbacks since then, so many fleeting victories, that Michael was wary of getting his hopes up. What he didn't know was that a clash unfolding deep inside the chemical industry was about to give him an unexpected advantage.

. . .

Because Michael's case named DuPont and 3M as defendants, his lawyers could subpoena and cross-check papers from various companies and piece together how some had collaborated to mislead the public. Similarly, attorneys for the other side mined the document cache for their own purposes. One of these lawyers was Amiel Gross, who oversaw PFAS litigation for Saint-Gobain.

Gross had been reviewing 3M papers when he made a disturbing discovery: Prior to the phaseout, his employer had been buying huge quantities of concentrated PFOA directly from the Minnesota company but hadn't factored it into its emissions calculations. The upshot was that Saint-Gobain's fabric-coating plants had actually spread far more PFOA than the company publicly admitted. This was no mere technicality—officials in some states were relying on these lowball figures to formulate their cleanup plans. Gross also discovered that Saint-Gobain had been using PFOA at several other plants across the country but hadn't informed regulators or investigated possible water pollution in surrounding communities.

Although Gross was a veteran corporate defense attorney, he'd spent most of his career in asbestos litigation, where the facts at play were historical and the executives who'd conspired to bury the awful truth were mostly dead. He'd always assumed these officials would have made different choices if they'd known that their machinations would wind up causing most of the domestic asbestos industry to collapse under the weight of its legal liabilities. Now, Gross had an opportunity to shape a similar scenario in real time.

So in early 2019, he met with the CEO of Saint-Gobain North America, Tom Kinisky, and proposed analyzing drinking water near all the company plants that had used PFOA but hadn't yet investigated the effect on nearby communities. The executive swiftly shut him down. "Don't do that," Kinisky warned, according to legal records. "If you look, you will find it. If you don't, you can say you didn't know."*[3]

Gross was floored. He believed that failing to investigate could open the company up to massive punitive damages—even criminal liability. It didn't make sense to him that the company would ignore these risks for the sake of quarterly profits. So he continued to press the matter with his superiors. According to Gross, he also urged them to disclose how much PFOA their fabric-coating plants in places like Hoosick Falls had actually spread.[4] Once again, he ran into a brick wall.

Gross eventually came up with a term to describe the phenomenon he was witnessing: It was a "generational passing of the buck" by the company's top leaders. "The mentality is, 'Why should I deal with the expense and the ramifications of this pollution when I can leave it to my successor?' " he explained. " 'By then, I'll be retired in the Hamptons.' "

The lawyer's tenacity eventually got him fired.[5] Under normal circumstances, this might have been the end of the story. Disputes between corporations and their lawyers rarely see the light of day, because of attorney-client privilege, not to mention the potential blowback for the lawyer's career. But on April 6, 2021, as Michael's lawyers were preparing for settlement talks, Gross lodged a complaint with the U.S. Department of Labor. Claiming that he'd been fired for whistleblowing, the attorney laid out a raft of allegations that could have been disastrous for Saint-Gobain if aired in court. Six days later, lawyers representing North Bennington residents filed an emergency motion to question Gross under oath. And after a series of rapid-fire negotiations, Saint-Gobain agreed to settle their case.[6]

---

* In a pair of written statements, Saint-Gobain dismissed Gross's claims about the company's operations, calling them "meritless," and disputed his account of the Kinisky meeting. "The conversation Mr. Gross alleges with former CEO Tom Kinisky never happened," the company asserted.

Then, in early May, Saint-Gobain, Honeywell, and 3M reached a settlement with Michael and other Hoosick Falls residents. The amount: a groundbreaking $65 million, far more per person than any class action in the history of forever-chemical litigation.

WHEN MICHAEL'S LAWYER CALLED to deliver the news, Michael felt like the wind had been knocked right out of him. It wasn't just the dollar figure that amazed him. For the first time in the history of New York State, his lawyer explained, the deal would fund a medical monitoring program to screen locals for the diseases associated with PFOA.* It also included a far more generous payout than similar cases. As compensation for the drop in property values, every village homeowner was entitled to roughly 10 percent of their home's assessed worth when the pollution became public. Those who relied on private wells at the time were eligible for an additional $10,000 or more, whether they owned or rented.[7]

As soon as Michael hung up the phone, he fired off a text to Marcus. Then he grabbed his father's golf clubs and headed to the green. Roaming the mostly empty course, he thought back on the wild journey of the past eight years: all the hurdles he'd overcome, all the experiences he never would have dreamed possible, all the friendships he'd formed with people who wouldn't have come into his orbit otherwise. Slowly it sank in that almost everything he'd been fighting for was actually coming to pass, and he was flooded with euphoria.

A few weeks later, Michael went out to celebrate at his favorite watering hole. He guzzled rum and Cokes and knocked back shots, taking stimulants to keep his energy up. When he finally arrived

---

* The settlement only guaranteed ten years' worth of medical monitoring. However, Michael's lawyers have continued pursuing their case against DuPont, which could yield funds to sustain the program longer.

home in the early hours of the morning, his heart started racing. Then, suddenly, his knees buckled and he crashed head-first into his coffee table. He came to in the emergency room, frightened and humiliated.

Again, Marcus pleaded with Michael to lay off the drinking and seek counseling. This time, Michael was shaken enough to agree, and soon he was speaking with a therapist every other week. Most of their sessions dealt with the guilt Michael had wrestled with ever since Ersel's death. He came to realize that continually retelling his father's story had kept his regret and anguish raw. Gradually, he began to let this burden go.

But Michael still had to contend with the suspicion of his fellow residents. When the settlement became public, he knew some people would see it as yet more proof he had used the water problems to enrich himself. It didn't matter that the entire village would benefit or that Michael's sole compensation would be the $25,000 awarded to lead plaintiffs for the work they'd put in. He also knew the rumors about his recent hospitalization would only fuel the questions about his character.

Hoping to keep ugly speculation to a minimum, Michael reached out to the Albany *Times Union* reporter who had broken the story of the water contamination back in 2015, and asked him to break the settlement story, too. He figured that if the facts were reported accurately and in enough depth, there would be less room for false assumptions.

The plan was for the article to go live as soon as the settlement papers were filed with the court on July 21, 2021. Early that afternoon, Michael arranged to have a *Times Union* photographer drop by his mother's house to take photos for the story. Michael's stomach was in knots. Sue, who had been in the hospital with serious lung problems, was ashen and emaciated. Removing her oxygen mask, she sat glumly next to Michael on the sofa, holding a picture of Ersel. "This might be a happy day for other people, but it sure as hell isn't for me," she told Michael after the photographer had left. "I'm living my life alone."

Michael spent the next few hours sprawled on his couch, swigging iced tea and obsessively hitting refresh on the *Times Union* home page. Finally, around 5 p.m., the story appeared.[8] Michael sat for a few moments in awed silence before posting it to Facebook with a message: "My father loved our community and giving back. This settlement in my opinion was him saving his biggest gift for last!" Just then, Emily burst into Michael's living room and gave him a big hug. They sat together on the sofa, their faces bathed in bluish light, as Michael scanned social media and answered a flurry of messages from people wanting to know how much money they'd be getting.

Afterward, Emily, Michael, and his girlfriend, Megan, relaxed in rocking chairs on the porch. As neighbors strolled by on their way to a community band concert in the park, fireflies flitted through the grass and "Amazing Grace" drifted toward them on the warm air.

By the time news of the settlement broke, the stigma of the contamination had finally faded and Hoosick Falls was experiencing something of a renaissance. Vibrant new businesses were springing up downtown. People from outside the area were moving in, drawn by the community's rustic charm. Houses that once would have languished on the market were fetching multiple above-asking offers.

News of the bidding wars got Emily thinking. Early in the pandemic, she'd taken nearly a year off to care for Ellie and help Gwen with school. At a time when many kids were struggling from lack of social contact, the girls flourished. But the loss of income had driven Emily even deeper into debt. She figured the best way to dig herself out was to sell her house and cash in on the appreciation.

Sure enough, when Emily put her house on the market in the spring of 2021, she immediately got an above-asking offer from a young woman who sent a letter saying she'd fallen in love with the place the moment she walked in. Emily was so touched that she refused to entertain any further bids. Even so, she wound up clearing enough to erase her debts and put aside $15,000. Emily, Jay, and the girls moved in with Emily's mother, who had recently bought a spacious home in Hoosick Falls. Emily hung a sign over the kitchen sink that read, "It Is What It Is."

In fact, while their new living situation wasn't exactly Cloud Nine, in some ways it turned out to be a wonderful arrangement. Emily enjoyed her mother's company, and Ellie thrived on her doting grandma's attention. Meanwhile, the family got some welcome news about Sahara, whose kidney disease had been causing serious complications—including two bouts of congestive heart failure. The hospital had identified an unrelated donor whose kidney was a match, and a transplant was in the works.

Then, in August of that year, Emily learned that Taconic Plastics in Petersburgh had agreed to settle her class action for $23.5 million. The terms were nearly identical to the Hoosick Falls deal, with even more generous financial compensation. Emily alone would get about $65,000 for property depreciation and "trespass" on her private well, plus fifteen years of medical monitoring for her family.[9] The money didn't feel like nearly enough to make up for the emotional damage her family had suffered, much less their health problems or their lifelong susceptibility to serious illness. Still, she was profoundly relieved that her years of fighting were finally over and that her children's health would be regularly monitored.

Four months later, New York State announced that it would move ahead with one of Emily's remaining priorities: a new water supply for Hoosick Falls. The plan was to tap into a separate, untainted aquifer *and* continue using the existing filters as an added safeguard. In another win for residents, Honeywell and Saint-Gobain had agreed to cover the entire $10 million cost and reimburse the state for the $35 million in taxpayer money spent dealing with the contamination.[10]

Now that local advocates had achieved almost all their goals, Emily began to rethink her activism. Some of her allies had joined a coalition seeking change on the national level, but Emily decided not to join them. "I don't want to miss any more of my children's lives," she explained. "I've missed too much already." Instead, she quit her job so she could spend the summer with her daughters. Come fall, Ellie would start kindergarten. Gwen would be a high school senior, ready to find her own way in the world. It was time to make up for the strain and heartache of the past six years, to replace those "shit memories with something beautiful."

But Michael wasn't done fighting. It had become painfully clear to him that many Hoosick Falls residents either didn't know about the settlement or didn't realize that they qualified. And few seemed to grasp just how big their payout could be.

Michael feared this lack of understanding would endanger the entire settlement since the companies were entitled to pull the plug if too many people opted out. So he organized two information sessions with his lawyers in Hoosick Falls, working doggedly to arrange everything, from the PA system to the folding chairs. But even after the events, the claims numbers barely budged. Michael knew from talking with elderly and working-class residents that one of the most daunting obstacles to filing claims was pulling together the necessary paperwork, including blood tests and water bills from 2016. So he approached local officials with a plan B: put on workshops where all the documents were easily accessible, with attorneys on hand to help.

The first session, in early December, was held in a dusty, wood-paneled corner of the armory. Two lawyers with laptops sat behind a plexiglass barrier, helping people file claims online and providing relevant records like water bills and property tax rolls. At another station, Health Department employees printed out blood-test results and well-sampling data. For the first hour or so, the crowd was sparse. But by early afternoon, the room was packed, and the harried lawyers had fallen so behind that people were waiting for hours. By the next session, though, the lawyers had worked out the kinks. Claims jumped from about 600 to 2,200 of the roughly 3,000 eligible residents—a participation rate that blew most class-action settlements out of the water.[11]

TAMI DUKET WAS DRIVING TO WORK on the morning of September 7, 2022, when her daughter called to say they'd received a $17,000 check from the water settlement. She was so stunned that she had to pull off the road. As a single mother working two jobs, Duket had sometimes struggled just to put food on the table. Now she'd be able to afford braces for her younger daughter and have enough left over to take both girls on the Disney World vacation they'd always dreamed of.

Similar scenes were playing out all over town as checks landed in mailboxes and people considered the possibilities this money could create. That afternoon, David Galusha, a lanky seventy-something resident, was padding around the gutted ground floor of his home in lambskin slippers, surveying the blistering wallpaper and bare studs. He planned to use his settlement money to fix it up so he and his wife could keep living in the house when they could no longer climb the rickety stairs to the finished second story.

West of town, Beatrice Berle was working on her organic farm when her $24,000 payment arrived. In recent years, the business had been straining under debt she'd taken on to modernize operations. Now, she realized, she could pay off the loan, ensuring the land she'd spent thirty years tending would continue providing for future generations.

Over the months that followed, signs of the windfall appeared everywhere as people renovated their houses and spruced up their storefronts. It seemed as if half the village was wearing new shoes or driving better cars. But once the initial excitement faded, few locals talked about the settlement or even admitted publicly that they had gotten any money. Finances were a private matter—and many people still didn't want to be seen as challenging Saint-Gobain.

Even fewer went out of their way to express appreciation toward Michael. Though he was disappointed, the slights didn't sting as much anymore. *He* knew he'd done everything in his power to protect his community. Now that his fight was over, Michael was sleeping better, and he wasn't tormented by nightmares about Ersel. Instead, he found himself reminiscing about the good times he'd had with his father.

Just before Christmas, Michael took Oliver in for their first round of medical monitoring. The program was run out of a former college gym west of Bennington, and offered free annual checkups and screening for linked conditions, plus special medical consultations for pregnant women. Thanks partly to the elaborate cash incentives provided by the settlement—participants got $100 after the first appointment and additional compensation if they stuck with the program over time—the response had been overwhelming.[12]

The moment Michael walked into the cavernous gym and saw a crowd of people waiting to check in, a wave of elation washed over him. *It was actually happening.* Michael and Oliver filled out their questionnaires and followed the line of traffic cones to the first station. A woman in a lab coat examined them and taught them how to check their testicles for lumps. A kindly gray-haired nurse took their blood and showed Oliver how she processed the samples, placing the vials in a spinning contraption that separated the red blood cells from the plasma. Afterward, she took Michael's hand and thanked him for fighting so hard to protect their community.

As he drove home through the Vermont woods, balsam fir and sugar maple streaking past his windows, Michael told his son the story of how the program came to be. A long time ago, when Oliver was just a toddler, Michael explained, he'd set out to ensure other people wouldn't have to suffer like Grandpa Ersel had. In the end he'd succeeded, not because he was especially smart, but because he refused to give up. Michael wanted Oliver to know that he could make a difference, too. It didn't matter if he was scraping by in school or whether he went on to have a high-powered job. "If you just work hard and surround yourself with the right people," Michael told him, "you can still accomplish great things."

19

To the Ends of the World

In early December 2021, a crowd descended on the Hoosick Falls Country Club to celebrate Marcus's fiftieth birthday. This wasn't a milestone Marcus had ever expected to see—for years, he'd been warning loved ones that the cancer would take him before then. And yet there he was, flashing his toothy smile as he chatted with old friends and danced to the Kool & the Gang classic "Get Down on It" with his ninety-something mother.

That he was able to do these things was something of a miracle: Earlier that year, his brain tumors had started growing again. The only real treatment option left was full-brain radiation, which can trigger steep cognitive decline. Marcus had opted to move ahead, and he seemed to emerge without any dire side effects.

In the months after the party, though, Marcus's family noticed that he was having memory problems. Sometimes, he would tee up a ball in the middle of a golf game, then forget where the ball was. Over the Fourth of July weekend, while Marcus was hanging out with some friends, he fell into a near-catatonic silence. Suspecting the problem was swelling in his brain, his doctors prescribed steroids, which made him more lucid. But from that point onward, he was a different person. He rarely expressed emotion or engaged in his usual lively banter, and he was no longer able to practice medicine.

Then, in the fall of 2022, Marcus's bone marrow suddenly stopped functioning, leaving his body unable to fight infection. After three weeks of fighting for his life in intensive care, he was able to go home, but he needed multiple weekly blood transfusions just to stay alive. And he was no longer eligible for other treatments to keep his cancer in check. As a result, by April 2023, the cancer had spread throughout his brain and his body.

There was only one thing Marcus really longed to do before he died. He was due to receive an award for his decades of devoted service to the community at a hospital gala in mid-June, and he desperately wanted to accept it in person. After that, he planned to stop taking the transfusions. Now, his wife, Gretchen, feared he wouldn't survive that long or wouldn't be able to grasp what was happening.

As the disease progressed, his memory was becoming more and more muddled. Sometimes he asked after people who had been dead for years. Most days, he'd spend hours on end watching TV reruns.

But somehow he retained his encyclopedic medical knowledge. When his nephew, Andrew, developed mysterious digestive problems, Marcus worked methodically to help him pin down the cause. "Maybe he couldn't tell you what he had for lunch," Andrew said. "But when we were in the moment talking about my symptoms, he would ask very detailed questions and say exactly what he thought the next steps should be." After Andrew's doctors misdiagnosed him with hemorrhoids, it was Marcus who recommended he see a particular specialist to undergo more testing. It turned out Andrew was actually suffering from ulcerative colitis—a PFOA-linked disease that causes irreparable damage if left untreated.

. . .

On the day of the gala, Marcus arrived in a crisp navy suit; Gretchen wore an elegant floor-length gown. When friends and colleagues flocked over to their table to greet them, many were taken aback by how much Marcus had changed. His face was swollen from the steroids, and he had trouble holding a conversation. But he gave out hugs and said a few heartfelt words to everyone.

After the dinner, when the hospital CEO Thomas Dee introduced his award, Marcus stood right up and headed to the stage before Dee could give his speech. Dee just smiled and delivered his remarks as the doctor stood beside him. Marcus was "one of the most beloved physicians" in the history of the hospital, Dee told the audience. "Marcus, I know I'm speaking for thousands when I say we are forever grateful for all that you've done," he continued, taking Marcus's hand. "And you know all of us in this room and the community love you."

With this, the crowd broke into a fervent standing ovation, but Marcus just stood there, wide-eyed. Some of the family members in attendance worried that the moment had been lost on him. But when Marcus finally returned to his seat, tears were streaming down his cheeks. Gretchen realized that he'd been absorbing everything that was happening and knew exactly what it meant.

Once Marcus stopped taking the transfusions, he sank deeper into his murky haze. He rarely left his armchair except to go to bed. Sometimes he spent hours just fiddling with his fingers. On Tuesday evenings, Michael would stay with him for a while to give Gretchen a break. Usually they'd get takeout and maybe chat about golf—one of the few topics besides medicine that engaged him. Occasionally, Michael tried to talk about the breakthroughs that were happening with the water issues, but Marcus couldn't seem to take it in.

Michael sometimes found himself counting the minutes until these visits were over. Ersel had been in terrible pain when he was dying, but at least he had been himself right up to the end. Marcus's cancer was smothering the very essence of who he was. Eventually, even his medical knowledge abandoned him. The process was unbearable for his loved ones to witness, but for Marcus, it was a kind of mercy. As Gretchen put it, "Marcus knew he was going to die soon. But that's where the thoughts stopped. He didn't dwell on it, because he didn't dwell on anything."

By the first of August, Marcus's body was shutting down. When Michael visited him that evening, he was lying unconscious in his bed. His skin was clammy, his breath labored. Michael sat for a while holding his dear friend's hand. Then he thanked Marcus for everything he'd done and told him he loved him—and for a brief instant, Marcus's eyes flickered open. Two day later, Marcus died, surrounded by friends and family.

In the weeks that followed, the entire village descended into mourning. Reading through the countless tributes on Facebook, Emily marveled at how many lives he'd touched: Practically everyone she knew, including people from outside the area, seemed to have a story of Marcus going to enormous lengths to help them.

Similarly, as articles about Marcus's fight for clean drinking water appeared in the regional media, many of his loved ones—Gretchen included—were surprised to learn he'd played such a crucial role.[1] Marcus had never really talked to them about his advocacy, just as he'd never sought credit for the extraordinary things he did for his patients. "He could go to the ends of the world for you, and it was just him doing his job," Andrew said. "That's how he felt."

AN HOUR BEFORE MARCUS'S WAKE, a line formed outside the funeral home. Soon it snaked down the block, past the faded clapboard rectory, and the half-empty parking lots, and Trustco Bank before disappearing down Elm Street. During the standing-room-only memorial the following day, the priest made a point of stepping down from the pulpit to be among his fellow mourners, in keeping with the humble way that Marcus had served his community. Laying his hand tenderly on the casket, he described the bitterly cold day in 2019 when they'd buried Marcus's father, Old Doc. The priest had fallen so ill with a fever that he had to rush home, and the doctor had left his grieving family to examine him and bring him medicine. Before departing, Marcus had leaned down and kissed the priest on the forehead, as he often did to comfort patients. "I knew at that moment that power of love," the priest told the audience. "Marcus does that for us. He does that for each one of us."

After the service, the hearse led a long procession of cars to the cemetery. As mourners lined up to lay gold and russet roses on the coffin, Michael hung back, lost in thought. It had been just over a decade since they'd buried Ersel in this same soil. At the time, Michael had been tormented by the sense that his father had been cheated, but he found it even more awful that Marcus had been taken so young. Not only had he been robbed of many years with his family; his loss had left a giant void at the heart of the community that could never be filled.

Michael felt sure that the people of Hoosick Falls would never again receive the kind of care Marcus had provided. No other doctor was going to come over at 3 a.m. to comfort a dying relative or show up at every high school basketball game just in case someone got injured. The old-fashioned doctoring that generations of Hoosick Falls residents relied on had died along with Marcus.

Still, Michael took comfort in knowing that Marcus's legacy would live on through their shared victories. Along with their friends and allies, they'd managed to secure a clean water supply for the future and ensure that every resident would have access to screening for the diseases caused by PFAS. Marcus wouldn't be around to witness all the lives that would be saved now that these illnesses could be detected early and treated. But in a very real way, Michael thought, his efforts would continue to safeguard the community's health for generations.

When the crowd finally dwindled, Michael laid a pale-yellow rose on Marcus's casket. Then he drove home, past the old brick factories and the colorful Victorian houses with American flags flapping from their porches—emblems of the community where he was born and where he planned to spend the rest of his days.

Epilogue

Right now, millions of people around the world are grappling with the question that lies at the heart of this book: How can they protect their families and communities from an insidious group of chemicals that permeates the bodies of all living beings from the moment of conception until death?

Almost every month brings bleak new revelations about how thoroughly forever chemicals have polluted the earth. One 2022 study of rainwater around the globe found that the levels of PFOA and PFOS alone had reached concentrations that endanger the health of people and ecosystems everywhere. Based on these facts, the authors concluded that the entire planet is now "outside the safe operating space" for humanity.[1] Surveys of rivers, lakes, drinking water, and food supplies around the world have yielded equally troubling findings.

Because these chemicals are so ubiquitous, even our well-meaning efforts to protect our health and environment can wind up increasing our exposure. Organic foods often contain harmful levels of PFAS for the same reason that animals in the wild do; they absorb these chemicals through ambient contamination of our soil and water. Many supposedly green products, like compostable food containers and "eco-friendly" children's clothing, have been shown to contain high concentrations, too.[2] Similarly, some of the places we think of as the most pristine are among the most heavily contaminated. Many small family-run farms throughout North America have levels of PFAS that rival the most polluted factory towns—the result of tainted sewage sludge being spread over their fields as fertilizer. A 2022 study of locally caught freshwater fish across the United States found that their flesh is so saturated with these chemicals that consuming even a single portion can increase our blood levels as much as drinking polluted water for an entire month.[3]

Moreover, the methods used to clean up PFAS pollution often end up returning the chemicals to the environment instead. The landfills where we bury forever-chemical waste, for instance, simply belch them back into the air. Trying to incinerate these substances, as many companies do, often just spreads them farther afield.[4]

Scientists once believed that our oceans might be our saving grace—that forever chemicals would wash out into the seas and be gradually diluted almost to the vanishing point. Unfortunately, the opposite is true. The chemicals latch on to tiny air bubbles that rise from ocean depths, eventually accumulating on the surface as a highly concentrated stew. When waves crash ashore, the bubbles burst, releasing immense quantities of PFAS into the atmosphere. As a result, scientists estimate, the oceans now emit more PFAS into the air than all of the world's industrial polluters combined.[5]

As disturbing as all these findings are, they encompass only a tiny fraction of the PFAS in the environment. To date, the scientific research on these chemicals has focused almost exclusively on two broad categories: legacy substances like PFOA, which contain eight or more carbon atoms and have been phased out in many countries; and short-chain molecules, like GenX, which contain between four and seven consecutive carbon atoms and are still widely used in manufacturing. In the past few years, researchers have realized that a third, barely studied category of forever chemicals might pose an even greater threat.

It all began in the early 2020s, when a group of European researchers developed a technique to detect an elusive subset of forever chemicals that evaded other methods—specifically, those with three or fewer carbon atoms. After applying this technique to drinking water across Germany, they made a breathtaking discovery: All but 2 percent of PFAS detected were ultrashort-chain substances, which up until that point had hardly registered on scientists' radar.

One molecule turned out to be particularly abundant—trifluoroacetic acid, or TFA, which is used to make pesticides, pharmaceuticals, and working fluids for heating and cooling systems. It is also a common breakdown product of other PFAS. The researchers found that this chemical alone accounted for 90 percent of forever chemicals detected in German tap water.[6] Since then, TFA has been detected at alarming levels in beer, bottled water, tea, and baby food in a variety of countries. One survey found that TFA made up virtually all the PFAS found in rivers and groundwater across Europe.[7]

Similarly, when researchers from Emory University measured the levels of various PFAS in tap water from homes in Indiana, they found that TFA accounted for 85 percent of the total. And the average concentration was orders of magnitude above the EPA's safety limit for PFOA. Industry has long insisted that shorter-chain chemicals are safer because they don't build up in people's bodies. But the Emory team found that the TFA levels in homeowners' blood were even higher than the national average for PFOA at its peak. This wasn't because the chemical had built up over time but because people were being exposed to such huge quantities.[8]

Since TFA has only recently attracted scientific scrutiny, data on its health effects are scarce, though animal studies have already linked it to familiar problems like liver damage and birth defects.[9] In other respects, ultra-short-chain chemicals like TFA are more worrisome than their better-studied cousins. They're even more mobile than short-chain chemicals like GenX. They build up faster in crops, leading to enormous concentrations in the few foods that have been tested. And they're virtually impossible to get out of drinking water. Not only do they pass right through the type of carbon filters used to remove legacy chemicals but they also foil newer treatment technologies that are meant to remove a broad range of PFAS.[10]

What's more, researchers warn that TFA "may only be the tip of the iceberg."[11] As detection methods improve, they are discovering more and more previously unknown PFAS. By the time they figure out where these substances come from or how they are affecting us, it will be too late: The planet may already be saturated with them. Meanwhile, manufacturers will have moved on to whole new sets of unknown molecules. Given the staggering scale of the problem, the burning question becomes, What can any of us do about it?

ALL ACROSS THE UNITED STATES, thousands of people whose lives and livelihoods have been upended by forever chemicals have opted for the same approach the Tennant family took in West Virginia: suing polluters. So far, manufacturers like DuPont, Chemours, and 3M have been hit with roughly fifteen thousand legal claims, the lion's share from municipalities, water districts, and residents of polluted communities, though more than thirty U.S. states have also brought cases. And the numbers are expected to rise steeply in the coming months and years. During an early 2024 plastic-industry conference, a prominent corporate lawyer warned executives to prepare for a "tsunami" of litigation that would "dwarf anything having to do with asbestos," one of the biggest and most costly legal fights in U.S. history. "Do what you can, while you can, before you get sued," he advised.[12]

Meanwhile, after decades of delays and half measures, regulators are finally taking steps to clamp down on forever chemicals. In April of 2024, the EPA issued binding standards for PFOA and PFOS in drinking water. Citing the mounting evidence that these chemicals damage the immune system at near-zero levels, the agency set the limits for each substance at 4 parts per trillion—the lowest concentration that can be reliably detected.[13]

While these standards are a potentially important step, they do nothing to protect people from the thousands of other forever chemicals that are inundating their bodies. And there's no telling whether they will survive Donald Trump's second presidency, though there is still strong bipartisan support for regulating PFAS in Congress.[14]

In the end, any federal action may prove less consequential than the aggressive measures cropping up in other places. In 2023, the European Commission introduced a wholesale ban on the production and sale of PFAS and products containing them, the most sweeping chemical regulation in the bloc's history. Thanks to the tireless efforts of grassroots activists, many U.S. states are embracing similar measures.

Just after European regulators made their move, Minnesota, the birthplace of commercial forever-chemical production, adopted a near-total ban on PFAS. It was named Amara's Law, after a young woman from the Saint Paul area, who grew up drinking water allegedly polluted by 3M and developed a rare form of liver cancer at age fifteen. She spent the final years of her life lobbying the state legislature to pass the bill without the loopholes sought by industry, which it did. Maine and Washington State have passed similar measures, and at least a dozen other states have enacted more-targeted bans on PFAS in various consumer products.[15]

Even before they take full effect, these regulations are proving to be genuinely game-changing. Suddenly, manufacturers in practically every industry are being forced to investigate their supply chains and provide regulators with detailed information about how they're using PFAS. For a growing number of companies, this information is such a mammoth PR liability that they are voluntarily migrating away from these substances. Dozens of major retailers with more than $700 billion in combined annual sales have already committed to eliminating forever chemicals from their packaging and products. These include Amazon, Starbucks, Apple, Target, McDonald's, and Home Depot.[16]

In a seismic shift from just a few years ago, some chemical manufacturers are abandoning PFAS, too. 3M, which owes more than $10 billion for PFAS settlements so far, has announced that it will quit producing the chemicals by the end of 2025. Among the reasons it cited were mounting regulation and pressure from investors who were troubled by the toll litigation had taken on the company's bottom line.[17]

3M is hardly the only chemical maker getting this kind of pushback: In 2023, a group of fifty-one global asset managers with more than $10 trillion in investment called on CEOs of the world's largest chemical companies to stop producing PFAS, warning they posed liability risks similar to the claims that all but wiped out the asbestos industry. The message seems to be getting through. Some industry leaders have signaled a willingness to phase out most of the forever chemicals on the market, given the right incentives.[18]

Rob Bilott, the lawyer who brought these chemicals to the world's attention, sees these developments as a testament to the power of ordinary citizens. "It shows just how much individual people and communities standing up and speaking out can do and the dramatic change they can put in motion," he said. "It took us way too long to get here, but it's happening."

And yet despite these hard-won concessions, the battle is far from over. Even as the chemical industry cedes ground in some areas, it is fighting ferociously to protect the most lucrative types of

PFAS: namely, fluoropolymers like Teflon and fluorinated gases, which together comprise the vast majority of the global PFAS market, with tens of billions of dollars in combined annual sales. Manufacturers argue that these substances—used in everything from rocket ships and air-conditioning to power lines—are indispensable to society and will only become more so, since they play a crucial role in green technologies like solar panels, heat pumps, and lithium-ion batteries.[19]

Led by Chemours, manufacturers have mounted a massive lobbying campaign to get these materials and their ingredients exempted from regulations. One of their chief targets is TFA, which is both a key component and a breakdown product of fluorinated gases. Already, the EPA office charged with vetting PFAS under the new toxic substances law has adopted an industry-backed definition that only includes chemicals with two or more adjoining fluorinated carbon atoms.[20] TFA has just one. As a result, the agency's analysis of the dangers posed by PFAS will completely ignore this substance, which is more abundant in the environment than all other forever chemicals combined.

In many ways, this situation is reminiscent of the global battle over another class of chemicals that consumed the world's attention during the 1980s and '90s: chlorofluorocarbons, or CFCs. Invented by the same GM chemists who developed leaded gasoline, CFCs first went on sale in the 1930s and quickly came to dominate the market for refrigerants. They later enabled innovations like home air-conditioning and played an integral role in the manufacturing of electronics. In the 1970s, however, University of California, Irvine, scientists discovered that as CFCs broke down in the atmosphere, they could deplete the ozone layer, which shields the earth from potentially lethal radiation. Alarmed, the researchers began advocating for a global ban on CFCs.

When Congress started to debate restrictions, DuPont responded with its usual playbook, organizing hundreds of CFC producers to discredit the science on ozone depletion. Together, they funded studies ostensibly proving that the ozone layer was fully intact. Regulators, the media, and even the scientific community took these results at face value, and for a while, the issue faded from public view.[21]

Then, in 1985, a British research team discovered a massive hole in the ozone layer over Antarctica. Suddenly, it was clear that ozone depletion was real, and that it was happening much faster than anyone had anticipated. Without steep cuts in CFC emissions, a National Resources Defense Council official warned, the world would soon face a crisis that would make "Chernobyl look like a trash fire at the county dump."[22]

The political reaction was swift and forceful. The Reagan EPA moved to ban the chemicals, and dozens of nations began negotiating a treaty to limit their use globally. Realizing restrictions were inevitable, DuPont adopted a strategy similar to the one the industry is now deploying in response to the rising tide of PFAS regulations. The company signaled support for the global agreement but insisted the targets were too aggressive. CFCs had many essential applications, DuPont argued; developing replacements would take years. In the end, delegates scaled back the targets. When the landmark Montreal Protocol was adopted in 1987, it only required signatories to cut CFC production by 50 percent over the next thirteen years.

But by that point, DuPont had begun to sense an opportunity. CFCs were a commodity product with multiple suppliers and relatively thin profit margins. Its substitutes would be proprietary substances that would command premium prices. Around the time the treaty was signed, the company began pouring resources into developing alternatives and quickly amassed more than twenty patents.

Once these were in hand, DuPont reversed course and began lobbying for a *faster* phaseout. Sure enough, by 1990, the Montreal Protocol had been amended to require a total elimination of all CFCs by 2000.[23] There was an exception for applications that were essential to public health and safety, or "critical for the functioning of society," but only when there were no "feasible alternatives." After the amendment was in place, substitutes flooded onto the market far faster than anyone had predicted, and levels of ozone-destroying molecules in the atmosphere fell sharply. As a result, the ozone layer is now on its way to a full recovery; the hole over Antarctica is expected to disappear entirely within the next forty years.[24]

ALL TOO OFTEN, we respond to grave environmental threats with a kind of collective paralysis. The problems are so vast and mind-bendingly complex that our individual efforts to address them can feel meaningless, especially when our political leaders are bent on rolling back protections. But the success of the Montreal Protocol proves that it is possible to solve even the most daunting environmental crises.

For years, scientists have been urging regulators to apply the lessons of the treaty by banning all but the most essential uses of PFAS, just as the protocol did with CFCs.[25] Some of the applications that the industry is maneuvering to protect would undoubtedly pass this test—lithium-ion batteries, for example, are crucial tools in the fight against climate change and can only be produced with the aid of fluoropolymers. Other uses clearly wouldn't qualify. Take fluorinated gases, the leading source of TFA in the environment. Most applications wouldn't be considered essential since there are already nontoxic, biodegradable alternatives available.

Many U.S. states are embracing this approach. Virtually all the recently enacted bans in places like Minnesota have included exceptions for applications that are "essential for health, safety, or the functioning of society." In 2024, the European Union followed suit and adopted the essential-use policy as one of its guiding principles for regulating hazardous chemicals generally.[26]

These are vital steps. But the Montreal Protocol also offers a more profound lesson about overcoming the enormous barriers to change that have been baked into our regulatory system ever since the battle over leaded gasoline in the 1920s. It was only when DuPont realized that it was in its own financial interest to stop fighting restrictions that the world was able to resolve the existential threat posed by CFCs. In that case it was the combination of global regulation, and the lucrative market for substitutes it opened, that persuaded the industry that abandoning CFCs was the most profitable option.

Something similar is starting to happen with PFAS, as the various regulations take effect in the United States and Europe and increasing numbers of consumers boycott products made with these chemicals. Already, entire industries that relied on forever chemicals are migrating away from them en masse. (Clothing manufacturing and fast-food restaurants are prime examples.) At the same time, companies from a wide variety of industries are racing to find substitutes for those applications where they don't yet exist.

These trends will only accelerate as the tidal wave of PFAS litigation engulfs consumer brands. L'Oréal has already been sued, as have the makers of Burt's Bees, Snapple, and Trojan condoms.[27] Because companies in virtually every sector of the global economy are vulnerable to similar litigation, forever-chemical manufacturers are coming under intense pressure from their corporate customers to develop safer alternatives. With so many varied interests clamoring for PFAS substitutes, the companies that bring them to market first stand to reap enormous profits—all of which pushes us closer to the kind of tipping point that led to the elimination of CFCs.

We have ordinary citizens to thank for these momentous shifts. Most of the improbable victories that have occurred so far were brought about by people in places like Hoosick Falls who dared to take on some of the world's most powerful corporations, often at great personal cost. People who fought to protect everything they held dear by filing lawsuits, holding protests, speaking out in the media, and demanding action from political leaders. Our best hope of stemming the devastating accumulation of these chemicals in the environment is to follow the trail they have blazed.

Notes

While researching this book, the author conducted more than six hundred interviews with hundreds of people on the front lines of the global forever-chemical crisis. Unless otherwise noted, the descriptions of events in contaminated communities are based on the author's firsthand observations or on interviews with those directly involved.

PREFACE

1. Andrea C. Gore et al., "Endocrine Disrupting Chemicals: Threats to Human Health," Endocrine Society and IPEN, February 2024; Jane Houlihan et al., "Body Burden: The Pollution in Newborns," Environmental Working Group, July 2005; Uloma Uche, "Pregnant with PFAS: The Threat of 'Forever Chemicals' in Cord Blood," Environmental Working Group, September 13, 2022; Rachel Morello-Frosch et al., "Environmental Chemicals in an Urban Population of Pregnant Women and Their Newborns from San Francisco," *Environmental Science & Technology* 50, no. 22 (October 2016): 12464–12472, doi:10.1021/acs.est.6b03492.
2. Gore, "Endocrine Disrupting Chemicals"; Gian Carlo Di Renzo et al., "International Federation of Gynecology and Obstetrics Opinion on Reproductive Health Impacts of Exposure to Toxic Environmental Chemicals," *International Journal of Gynecology and Obstetrics* 131, no. 3 (December 2015): 219–225, doi:10.1016/j.ijgo.2015.09.002.
3. At the time, Deer Park said it planned to phase BPA out of its five-gallon

refillable water bottles within two years. According to Deer Park's new parent company, BlueTriton Brands, the containers are now BPA-free.

4. Specifically, the plaintiff, Eastman Chemical Co., accused the scientists of infringing on its trademark by publicly alleging that the company's popular Tritan plastic—which was advertised as free of estrogen-disrupting chemicals—was actually a potent estrogen mimicker. The case centered on the business dealings of the study's coauthor, George Bittner, a professor of neurobiology at the University of Texas. He also owned a company called PlastiPure that developed nonestrogenic plastics. Based on this, Eastman argued that Bittner's team had engaged in false advertising and unfair competition by spreading damaging information about its products. Tellingly, documents released through litigation revealed that Eastman's own studies had found that one of Tritan's ingredients was likely *more* estrogenic than BPA. Nevertheless, the jury decided in the company's favor and barred the defendants from publicly stating that Tritan was estrogenic, except in published scientific studies. Chun Z. Yang et al., "Most Plastic Products Release Estrogenic Chemicals: A Potential Health Problem That Can Be Solved," *Environmental Health Perspectives* 119, no. 7 (July 2011), doi: 10.1289/ehp.1003220; Eastman Chemical Co. v. PlastiPure, Inc., no. 775 F.3d 230 (5th Circuit 2014).
5. Mariah Blake, "The Scary New Evidence on BPA-Free Plastics," *Mother Jones*, March/April 2014.
6. **On polar bears:** Gabriel Boisvert et al., "Bioaccumulation and Biomagnification of Perfluoroalkyl Acids and Precursors in East Greenland Polar Bears and Their Ringed Seal Prey," *Environmental Pollution* 252, pt. B (September 2019): 1335–1343, doi:10.1016/j.envpol.2019.06.035. **On eagles:** K. Kannan et al., "Perfluorooctane Sulfonate in Fish-Eating Water Birds Including Bald Eagles and Albatrosses," *Environmental Science & Technology* 35, no. 15 (August 2001): 3065–3070, doi:10.1021/es001935i. **On fish:** Daniele de A. Miranda et al., "Perfluoroalkyl Substances in the Western Tropical Atlantic Ocean," *Environmental Science & Technology* 55, no. 20 (October 2021): 13749–13758, doi:10.1021/acs.est.1c01794. **On Mount Everest:** K. R. Miner et al., "Deposition of PFAS 'Forever Chemicals' on Mt. Everest," *Science of the Total Environment* 759, art. 144421 (March 2021), doi:10.1016/j.scitotenv.2020.144421. **On rain:** Ian T. Cousins et al., "Outside the Safe Operating Space of a New Planetary Boundary for Per- and Polyfluoroalkyl Substances (PFAS)," *Environmental Science & Technology* 56, no. 16 (August 2022): 11172–11179, doi:10.1021/acs.est.2c02765.
7. See, for example, K. Kannan et al., "Perfluorooctanesulfonate and Related Fluorochemicals in Human Blood from Several Countries," *Environmental Science & Technology* 38, no. 17 (September 2004): 4489–4495, doi:10.1021/es0493446; Andrew B. Lindstrom et al., "Polyfluorinated Compounds: Past, Present, and Future," *Environmental Science & Technology* 45, no. 19 (September 2011): 7954–7961, doi:10.1021/es2011622.
8. Kyle Steenland et al., "Review: Evolution of Evidence on PFOA and Health Following the Assessments of the C8 Science Panel," *Environment Interna-*

tional 145, art. 106125 (September 2020), doi:10.1016/j.envint.2020.106125; U.S. Environmental Protection Agency, "Our Current Understanding of the Human Health and Environmental Risks of PFAS," updated May 16, 2024, https://www.epa.gov/pfas/our-current-understanding-human-health-and-environmental-risks-pfas.

9. Mariah Blake, "Welcome to Beautiful Parkersburg, West Virginia," *HuffPost Highline,* August 27, 2015; Tennant v. E.I. du Pont de Nemours & Co., No. 6:99-0488 (S.D. W.Va. 1998).
10. There was also another factor driving the surge in public awareness: Between August 2015 and January 2016, three national news outlets ran major stories about the Parkersburg saga. See Blake, "Beautiful Parkersburg, West Virginia"; Sharon Lerner, "The Teflon Toxin," *The Intercept,* August 11, 2015; Nathaniel Rich, "The Lawyer Who Became DuPont's Worst Nightmare," *New York Times Magazine,* January 6, 2016.

CHAPTER 1: A BUMP IN THE ROAD

1. The mumbling is a side effect of hearing problems Michael developed as a result of his frequent childhood ear infections.

CHAPTER 2: TEFLON TOWN

1. "Art: Presents from Grandma," *Time,* December 28, 1953; "Grandma Moses Is Dead at 101; Primitive Artist 'Just Wore Out,'" *New York Times,* December 14, 1961.
2. Anna Mary Robertson Moses, *Grandma Moses: My Life's History* (New York: Harper & Brothers, 1948), 129–130; Katherine Jentleson and Jane Kallir, "From Gallery to Greeting Card: Copyrighting Grandma Moses," *Art Journal* 76, no. 1 (July 2017): 90–94; "Six Women Named for Press Awards," *New York Times,* April 16, 1949.
3. "Art: Presents from Grandma."
4. George Baker Anderson, *Landmarks of Rensselaer County* (Syracuse: D. Mason & Company, 1897), 660–662; *Standard Directory, 1904–1905* (Hoosick Falls, NY: D. L. Hall, 1904); "Wood Company Changes Plans: Farm Implement Company Discontinues Making This Kind of Machinery—Affects Local Dealers," *Star-Gazette* (Elmira, NY), January 6, 1923.
5. "Cleveland H. Dodge, Philanthropist, Dies," *New York Times,* June 25, 1926.
6. Mike McGonnigal, "Cleveland Dodge Spurred Rebirth of Area Village," *Sunday Record* (Troy, NY), April 10, 1980.
7. Philip Leonard, "Short History of Dodge Industries 1955–1967," Hoosick Township Historic Society, June 2004; "Pownal Man Heads Business Firm in Hoosick Falls," *Bennington (VT) Banner,* October 18, 1955; Obituary for Phyllis Boushall Dodge, *Bennington (VT) Banner,* January 20, 2004.

8. Leonard, "Short History of Dodge Industries"; "Work Starts on Fiber Corp. $500,000 Plant Addition," *Troy (NY) Record,* August 4, 1966; "Dodge Fibers: After 11 Years, a Name Change," *Bennington (VT) Banner,* December 8, 1966; "Dodge Fibers Names Export Manager," *Troy (NY) Record,* January 29, 1965.
9. "Dodge Fibers: After 11 Years, a Name Change."
10. David Scribner, "Chemical Coating Firm to Open Here," *Bennington (VT) Banner,* November 23, 1968; Matilda McQuaid, *Extreme Textiles: Designing for High Performance* (Princeton, NJ: Princeton Architectural Press, 2005), 147.
11. Obituary for Lester Russell, *Bennington (VT) Banner,* September 12, 2008.
12. "Hoosick Falls Mayor Cites Growth, Lists Program," *Troy (NY) Record,* January 27, 1967.
13. The firm was acquired by Saint-Gobain in 1999, shortly before Michael graduated from college.

CHAPTER 3: LUCIFER'S GAS

1. Roy J. Plunkett, transcripts of interviews by James J. Bohning, New York City, April 14, 1986, and Philadelphia, May 27, 1986, Science History Institute, Oral History Program, Philadelphia (hereafter cited as Plunkett, Science History Institute Interviews).
2. Milton Sherman, "Taming Chemistry's Hellcat," *Collier's Weekly,* February 19, 1949; R. E. Banks, D. W. A. Sharp, and J. C. Tatlow, *Fluorine: The First Hundred Years (1886–1986),* (Switzerland: Elsevier, 1986).
3. Specifically, Plunkett's colleague was charged with developing an alternative to the most sought-after variant, Freon-114, which DuPont produced through Kinetic Chemicals, a joint venture with GM. See "Suit Shows DuPont-GM Clash over Freon-114 Sale," *Refrigeration Engineering,* vol. 61 (Washington, DC: U.S. Office of Technical Services, 1953), 535; Anne Cooper Funderburg, "Making Teflon Stick," *American Heritage's Invention & Technology* 16, no. 1 (fall 2000).
4. Plunkett, Science History Institute Interviews; Roy J. Plunkett, "The History of Polytetrafluoroethylene: Discovery and Development," in *High Performance Polymers: Their Origin and Development,* ed. Raymond B. Seymour and Gerald S. Kirshenbaum (Dordrecht: Springer, 1986): 261–266.
5. Plunkett, Science History Institute Interviews; Funderburg, "Making Teflon Stick."
6. Roy J. Plunkett's laboratory notebook, Hagley Museum and Library, Wilmington, DE; Roy J. Plunkett, Tetrafluoroethylene Polymers, US Patent 2,230,654 A, filed July 1, 1939, issued February 4, 1941; Plunkett, Science History Institute Interviews.
7. Robert F. Haggard, "The Politics of Friendship: Du Pont, Jefferson, Madison,

and the Physiocratic Dream for the New World," *Proceedings of the American Philosophical Society* 153, no. 4 (December 2009): 419–440.

8. Thomas Jefferson to Pierre Samuel du Pont de Nemours, January 18, 1802, Jefferson Papers, *Founders Online*, National Archives (hereafter cited as Jefferson Papers).
9. Marc Duke, *The Du Ponts: Portrait of a Dynasty* (New York: Saturday Review Press, 1976), 50–51, 58-59; Gerard Colby, *Du Pont Dynasty: Behind the Nylon Curtain* (New York: Open Road Media, 2014), 11, 95–98.
10. Duke, *Du Ponts: Portrait of a Dynasty,* 80–83; American Society of Mechanical Engineers International, "Brandywine River Powder Mills," October 9, 2002.
11. Éleuthère Irénée du Pont de Nemours to Thomas Jefferson, July 20, 1803, Jefferson Papers; Thomas Jefferson to Henry Dearborn, July 29, 1803, Jefferson Papers.
12. Alfred D. Chandler and Stephen Salsbury, *Pierre S. Du Pont and the Making of the Modern Corporation* (New York: Harper & Row, 1971), 57–61, 69, 72; Mark R. Wilson, "Gentlemanly Price-Fixing and Its Limits: Collusion and Competition in the U.S. Explosives Industry During the Civil War Era," *Business History Review* 77, no. 2 (2003): 207–234, doi.org/10.2307/30041144.
13. The $3 million figure is based on corporate earnings estimates from 1906. Chandler and Salsbury, *Pierre S. Du Pont,* 267; Colby, *Behind the Nylon Curtain,* 147–148; U.S. Congress, Senate Special Committee to Investigate the Munitions Industry, *Munitions Industry: Hearings Before the Special Committee Investigating the Munitions Industry,* 73rd Cong. (1935), 4207–4387.
14. Chandler and Salsbury, *Pierre S. Du Pont,* 48–54, 55–93, 124–125, 147; David A. Hounshell, "Measuring the Return on Investment in R&D: Voices from the Past, Visions of the Future," in *Assessing the Value of Research in the Chemical Sciences: Report of a Workshop,* ed. Chemical Sciences Roundtable, National Research Council (Washington, DC: National Academies Press, 1998).
15. The White House, "Theodore Roosevelt: The 26th President of the United States," courtesy of the White House Historical Association, accessed in August 2024, www.whitehouse.gov/about-the-white-house/presidents/theodore-roosevelt/.
16. Chandler and Salsbury, *Pierre S. Du Pont,* 261.
17. "The Powder Octopus," *National Tribune* (Washington, DC), January 17, 1907; "Independents Attack the Powder Trust," *New York Times,* January 24, 1907; "Powder Trust Has Grip on Uncle Sam," *Indianapolis News,* March 8, 1906; "Nation in Grip of Powder Trust," *Chicago Daily Tribune,* February 24, 1906.
18. "Power Trust Suit Papers Are Filed," *New York Times,* July 13, 1907; Chandler and Salsbury, *Pierre S. Du Pont,* 262–271, 274–282.
19. David A. Hounshell and John Kenly Smith, Jr., *Science and Corporate Strategy: Du Pont R&D, 1902–1980* (Cambridge: Cambridge University Press, 1988), 13, 29–32, 54–55, 57, 65–66.

20. Final Decree, United States v. E.I. du Pont de Nemours & Co, 188 F. 127 (3d Cir. 1911), June 13, 1912.
21. Chandler and Salsbury, *Pierre S. Du Pont,* 289–290.
22. Specifically, DuPont's profits in 1916 were more than ten times the annual average in the decade leading up to the war. Its gains for the remainder of the conflict were slightly lower than in 1916 but far higher than in any previous era. See Chandler and Salsbury, *Pierre S. Du Pont,* 359–360.
23. Hounshell and Smith, *Science and Corporate Strategy,* 78–96, 132.
24. Hounshell and Smith, *Science and Corporate Strategy,* 77, 87–88.
25. Hounshell and Smith, *Science and Corporate Strategy,* 82–83.
26. Alexander Mitchell Palmer, "Development of American Dye Industry," in *1919 Year Book of the Oil, Paint and Drug Reporter,* comp. William O. Allison (New York: Oil, Paint and Drug Reporter Inc., 1920).
27. Kathryn Steen, "Patents, Patriotism, and 'Skilled in the Art': USA v. The Chemical Foundation, Inc., 1923–1926," *Isis* 92, no. 1 (March 2001): 91–122.
28. Palmer, "American Dye Industry."
29. Hounshell and Smith, *Science and Corporate Strategy,* 92–93.
30. "Seek to Arrest German Chemists: Warrants Issued at Cologne for Four Men Alleged to Have Illegal Contract with Du Ponts in America," *Burlington (VT) Free Press,* February 21, 1921; Hounshell and Smith, *Science and Corporate Strategy,* 92–93; Matthew E. Hermes, *Enough for One Lifetime: Wallace Carothers, Inventor of Nylon* (Washington, DC: American Chemical Society and the Chemical Heritage Foundation, 1996), 61.
31. Joseph Borkin, *The Crime and Punishment of I. G. Farben* (New York: Macmillan, 1978), 39.
32. Durant's ouster followed a lengthy campaign by Pierre DuPont's camp to rein in Durant's impulsive management style. In 1920, the stock market crashed and the share price for GM, which was in the midst of a costly, haphazard Durant-led expansion, plummeted. Durant tried to break the fall by buying huge blocks of shares, but the plan backfired. By that November, he had $30 million in debt secured by GM stock, and the margins were rapidly dwindling. Suspecting the entire company was on the verge of collapse, Pierre and his GM allies held an emergency meeting with Durant's banker, a partner at J. P. Morgan & Company, which confirmed Pierre's fears. So the men came up with a plan to bail Durant out, in return for his resignation and the bulk of his GM stock. See Chandler and Salsbury, *Pierre S. Du Pont,* 435–480.
33. Chandler and Salsbury, *Pierre S. Du Pont,* 482–500; David Farber, *Sloan Rules: Alfred P. Sloan and the Triumph of General Motors* (Chicago: University of Chicago Press, 2002), 82.
34. Charles Franklin Kettering, *Biographical Memoir of Thomas Midgley, Jr., 1889–1944* (Washington, DC: National Academy of Sciences, 1947); Jamie Lincoln Kitman, "The Secret History of Lead," *The Nation,* March 2, 2000; Mary C. Rabbitt, *The United States Geological Survey, 1879–1989* (Washington, DC: Government Printing Office, 1989), 28.

35. Hounshell and Smith, *Science and Corporate Strategy,* 150–151; Kitman, "Secret History of Lead."
36. Hounshell and Smith, *Science and Corporate Strategy,* 151–152; Silas Bent, "Tetraethyl Lead Fatal to Makers," *New York Times,* June 22, 1925.
37. Specifically, the two Wilmington (DE) papers, *The Morning News* and *The Evening Journal,* were owned by Christiana Securities, a holding company controlled by DuPont. See Ben Bagdikian, "Case History: Wilmington's 'Independent' Newspapers," *Columbia Journalism Review,* Summer 1964.
38. Gerald Markowitz and David Rosner, *Deceit and Denial: The Deadly Politics of Industrial Pollution* (Berkeley: University of California Press, 2002), 20–21; Rob Rapley, producer and director, and Michelle Ferrari, writer, "The Poisoner's Handbook," *American Experience,* PBS, aired December 2, 2013.
39. Christian Warren, *Brush with Death: A Social History of Lead Poisoning* (Baltimore: Johns Hopkins University Press, 2001), 124–125; "Many Near Death as Result of Inhaling 'Looney Gas' Fumes," *Columbus (NE) Telegram,* October 30, 1924.
40. "Sees Deadly Gas a Peril in Streets: Dr. Henderson Warns Public Against Auto Exhaust of Tetra-Ethyl Lead," *New York Times,* April 22, 1925.
41. Farber, *Sloan Rules,* 84; Kitman, "Secret History of Lead"; U.S. Public Health Service, "Proceedings of a Conference to Determine Whether or Not There Is a Public Health Question in the Manufacture, Distribution, or Use of Tetraethyl Lead Gasoline," *Public Health Bulletin* 158 (Washington, DC: Government Printing Office, August 1925), 13–20.
42. Robert A. Kehoe, "Review of *Silent Spring,* by Rachel Carson," prepared for *Farm Quarterly,* 1963, Robert A. Kehoe Archival Collection, University of Cincinnati, Health Sciences Library, Henry R. Winkler Center for the History of the Health Professions (hereafter cited as Kehoe Archives).
43. Devra Davis, *The Secret History of the War on Cancer* (New York: Basic Books, 2007), 79–80.
44. U.S. Public Health Service, "Conference to Determine Whether or Not There Is a Public Health Question," 17–20; Robert A. Kehoe et al., "A Study of the Health Hazards Associated with the Distribution and Use of Ethyl Gasoline," Eichberg Laboratory of Physiology, University of Cincinnati, April 1928, accessed via Toxic Docs, www.toxicdocs.com.
45. Markowitz and Rosner, *Deceit and Denial,* 27–29.
46. U.S. Public Health Service, "Conference to Determine Whether or Not There Is a Public Health Question," 70.
47. David Rosner and Gerald Markowitz, "A 'Gift of God'? The Public Health Controversy over Leaded Gasoline During the 1920s," *American Journal of Public Health* 75, no. 4 (April 1985): 344–352, doi:10.2105/ajph.75.4.344.
48. Jerome O. Nriagu, "Clair Patterson and Robert Kehoe's Paradigm of 'Show Me the Data' on Environmental Lead Poisoning," *Environmental Research* 78, no. 2 (August 1998): 71–78, doi:10.1006/enrs.1997.3808.
49. "Gift Is Accepted by the Board for Construction and Maintenance of College

of Medicine Laboratory—Donation Is $130,000," *Cincinnati Enquirer,* October 8, 1929.

50. Nriagu, "Clair Patterson and Robert Kehoe's Paradigm"; Kitman, "Secret History of Lead."
51. Chandler and Salsbury, *Pierre S. Du Pont,* 573.
52. Hounshell and Smith, *Science and Corporate Strategy,* 226.
53. Charles M. A. Stine, "Molders of a Better Destiny," *Science* 96, no. 2492 (October 1942): 305–311, doi:10.1126/science.96.2492.305.
54. Hounshell and Smith, *Science and Corporate Strategy,* 223–225, 233.
55. Hermes, *Enough for One Lifetime,* 77–84; Sam Knight, "The Tragic Story of Wallace Hume Carothers," *Financial Times,* November 28, 2008.
56. American Chemical Society, "The Establishment of Modern Polymer Science by Wallace Carothers," commemorative booklet, National Historic Chemical Landmarks Program, November 17, 2000; Hermes, *Enough for One Lifetime,* 89–93; Knight, "Tragic Story of Wallace Hume Carothers."
57. Notably, although Bakelite was discovered by accident, it took months of research and experimentation by the renowned chemist Leo Baekeland to transform it into a viable product. National Museum of American History and the American Chemical Society, "The Bakelizer," November 9, 1993.
58. Yasu Furukawa, *Inventing Polymer Science: Staudinger, Carothers, and the Emergence of Macromolecular Chemistry* (Philadelphia: University of Pennsylvania Press, 1998), 75; Knight, "Tragic Story of Wallace Hume Carothers."
59. Hermes, *Enough for One Lifetime,* 91.
60. Hermes, *Enough for One Lifetime,* 112–115; Knight, "Tragic Story of Wallace Hume Carothers."
61. Hermes, *Enough for One Lifetime,* 151–152.
62. Hounshell and Smith, *Science and Corporate Strategy,* 243–246; Knight, "Tragic Story of Wallace Hume Carothers"; Hermes, *Enough for One Lifetime,* 157–159, 182–189, 285–291.
63. H. C. Engelbrecht and F. C. Hanighen, *Merchants of Death: A Study of the International Traffic in Arms* (New York: Dodd, Mead and Co., 1934).
64. Jeffrey L. Meikle, *American Plastic: A Cultural History* (New Brunswick, NJ: Rutgers University Press, 1997), 133–140; Robert H. Ferrell, "The Merchants of Death, Then and Now," *Journal of International Affairs* 26, no. 1 (1972): 29–39.
65. The committee conducting the investigation ultimately concluded that there was some merit to these allegations. As it reported to Congress, "There is no question that these attempts were discussed, were planned, and might have been placed in execution when and if the financial backers deemed it expedient." See Sally Denton, *The Plots Against the President: FDR, a Nation in Crisis, and the Rise of the American Right* (London: Bloomsbury Press, 2012), 208.
66. Meikle, *American Plastic,* 132–133; William L. Bird, *Better Living: Advertising, Media and the New Vocabulary of Business Leadership, 1935–1955* (Evanston, IL: Northwestern University Press, 1999).

67. New York Herald Tribune, *America Facing Tomorrow's World: Report of the Eighth Annual New York Herald Tribune Forum on Current Problems: The Waldorf-Astoria, October 25 and 26, 1938, the New York World's Fair, October 27, 1938* (New York: New York Herald Tribune, 1938); Meikle, *American Plastic,* 133–144.
68. American Chemical Society, "Establishment of Modern Polymer Science by Wallace Carothers."
69. Plunkett, Science History Institute Interviews.
70. W. E. Hanford and R. M. Joyce, "Polytetrafluoroethylene," *Journal of the American Chemical Society* 68, no. 10 (October 1946): 2082–2085; Plunkett, Science History Institute Interviews; Malcolm M. Renfrew, transcript of interview by James Bohning, New Orleans, August 31, 1987, Science History Institute, Oral History Program, Philadelphia (hereafter cited as Renfrew, Science History Institute Interview).

CHAPTER 4: EXILE TO DEVIL'S ISLAND

1. William Lanouette and Bela Szilard, *Genius in the Shadows: A Biography of Leo Szilard, the Man Behind the Bomb* (Chicago: University of Chicago Press, 1994), 235–236; Spencer Rumsey, "Einstein's Long Island Summer of '39," *Long Island Press,* February 1, 2013.
2. Lanouette and Szilard, *Genius in the Shadows,* 198–199, 235–236.
3. Lanouette and Szilard, *Genius in the Shadows,* 234–244; Albert Einstein to F. D. Roosevelt, President of the United States, August 2, 1939, Franklin D. Roosevelt Presidential Library and Museum, Hyde Park, NY.
4. F. G. Gosling, "The Manhattan Project: Making the Atomic Bomb," U.S. Department of Energy, Office of History and Heritage Resources, January 2010, 6–7.
5. U.S. Department of Energy, "Safety Data Sheet Uranium Hexafluoride (UF6)," created March 18, 1994, revised June 23, 2020; Richard Rhodes, *The Making of the Atomic Bomb* (New York: Simon & Schuster, 1988), 494–495.
6. H. Goldwhite, "The Manhattan Project," *Journal of Fluorine Chemistry* 33, nos. 1–4 (1986): 113; James Barron, "Joseph H. Simons Dies at 86; Pioneer in Fluorocarbon Use," *New York Times,* January 3, 1984.
7. Renfrew, Science History Institute Interview.
8. Renfrew, Science History Institute Interview.
9. William E. Hanford, transcript of interview by James J. Bohning, Bethesda, March 15, 1995, Science History Institute, Oral History Program, Philadelphia.
10. R. E. Banks, B. E. Smart, and J. C. Tatlow, eds., *Organofluorine Chemistry: Principles and Commercial Applications* (New York: Springer, 1994), 16.
11. U.S. Atomic Energy Commission, *Manhattan District History,* book VII, *Feed Materials, Special Procurement, and Geographical Exploration,* vol. 1, *Feed Materials and Special Procurement* (Washington, DC: U.S. Atomic Energy

Commission, June 12, 1947) (hereafter cited as AEC, *Manhattan District History*, book VII, vol. 1), G.24; Goldwhite, "The Manhattan Project," 109–132.

12. After receiving several messages and an unannounced visit from other chemists who purported to be doing similar work, one Cornell professor who was developing fluorocarbons under government contract informed his handler that he was "disturbed by this apparent lack of cooperation between different groups in the Government service." See William T. Miller to C. S. Marvel, Department of Chemistry at University of Illinois, March 11, 1942, William T. Miller Papers, Division of Rare and Manuscript Collections, Cornell University Library (hereafter cited as Miller Papers).
13. Minutes from a conference on fluorocarbons, Dumbarton Oaks, Washington, DC, July 17, 1942, Miller Papers.
14. Jonathan W. Williams, Office for Emergency Management, National Defense Research Committee of the Office of Scientific Research and Development, to William T. Miller, Jr., July 20, 1943, Miller Papers; AEC, *Manhattan District History*, book VII, vol. 1, G.31–32.
15. AEC, *Manhattan District History*, book VII, vol. 1, G.3, G.30.
16. Gosling, "Manhattan Project: Making the Atomic Bomb," 5–14; Rhodes, *Making of the Atomic Bomb*, 425–425; Laurence Lippsett, "The Race to Make the Bomb: The Manhattan Project; Columbia's Wartime Secret," *Columbia College Today*, spring/summer 1995.
17. Alex Wellerstein, "Remembering the Chicago Pile, the World's First Nuclear Reactor," *New Yorker*, December 2, 2017; Steve Koppes, "How the First Chain Reaction Changed Science," *University of Chicago News*, December 10, 2012.
18. Leslie R. Groves, *Now It Can Be Told: The Story of the Manhattan Project* (Boston: De Capo Press, 2009), 37.
19. Groves shared this highly confidential information with Stine and his colleague although they lacked security clearance. According to Groves's memoir, Manhattan Project leadership routinely did this when trying to enlist corporations: "We talked to one or two responsible officials, preferably men already known to us, whose judgment and security-mindedness we had no reason to doubt. The urgency of the project did not allow time for us to conduct any detailed security checks in advance of negotiations; instead, we relied upon the discretion and patriotism of American industry." See Groves, *Now It Can Be Told*, 59.
20. Groves, *Now It Can be Told*, 58.
21. Groves, *Now It Can Be Told*, 59–61.
22. General Kenneth Nichol, transcript of interview by Stephane Groueff, January 5, 1965, Stephane Groueff Collection, Howard Gotlieb Archival Research Center, Boston University.
23. Frank L. Kluckhohn, "Arnold Says Standard Oil Gave Nazis Rubber Process," *New York Times*, March 27, 1942; Kitman, "Secret History of Lead."
24. Groves, *Now It Can Be Told*, 62–63.
25. Crawford Greenewalt, transcript of interview by Stephane Groueff, Wilmington, September 12, 1965, Stephane Groueff Collection, Howard Gotlieb Ar-

chival Research Center, Boston University (hereafter cited as Greenewalt, Stephane Groueff Interview).

26. Tom Gary, transcript of interview by Stephane Groueff, September 12, 1965, Stephane Groueff Collection, Howard Gotlieb Archival Research Center, Boston University.
27. Compton's letter was technically to Conant, but Compton sent a copy directly to Groves with a personal message: "This letter is addressed to Conant, but I am equally anxious that you give it immediate and careful attention." Arthur H. Compton to Dr. James B. Conant, November 23, 1942, Bush-Conant Files Relating to the Development of the Atomic Bomb, 1940–1945, National Archives and Records Administration (hereafter cited as Bush-Conant Files); Groves, *Now It Can Be Told,* 45–46; "Time Table," September 9, 1943, Hagley Library and Archives, Wilmington, DE: E.I. du Pont de Nemours & Company Atomic Energy Division, including Clinton, Hanford, and Savannah River administrative records, accession 1957, box 1, folder 2; Greenewalt, Stephane Groueff Interview.
28. Lanouette and Szilard, *Genius in the Shadows,* 280–282.
29. Crawford Greenewalt, "Unclassified Portions of C. H. Greenewalt's Notes," vol. 3, Hagley Library and Archives, Wilmington, DE, accession 1889, 1942.
30. W. K. Lewis to General L. R. Groves, Washington, DC, "Lewis Report," December 7, 1942, Papers of Lt. Gen. Leslie R. Groves, Jr., National Archives and Records Administration, (hereafter cited as Groves Papers); Gosling, "Manhattan Project: Making the Atomic Bomb," 16.
31. Gosling, "Manhattan Project: Making the Atomic Bomb," 16–17.
32. Arthur H. Compton Metallurgical Laboratory, University of Chicago, to Eugene P. Wigner, "Necessity for Cooperation with DuPont," memorandum, July 23, 1943, Bush-Conant Files; Andrew Szanton, *The Recollections of Eugene P. Wigner as Told to Andrew Szanton* (New York: Plenum Press, 1992), 233–235.
33. For more on the plutonium patents, see Groves, *Now It Can Be Told,* 47; Gosling, "The Manhattan Project: Making the Atomic Bomb," 31. For the Teflon patent, see Roy J. Plunkett, Tetrafluoroethylene Polymers, US Patent 2,230,654 A, filed July 1, 1939, issued February 4, 1941; DuPont's District Engineer Ruhoff to Dr. H. T. Wensel, Clinton Engineer Works, March 30, 1944, Christopher Bryson and Joel Griffiths Papers, Special Collections and University Archives, University of Massachusetts Amherst (hereafter cited as Bryson and Griffiths Papers).
34. U.S. Department of Energy, Office of History and Heritage Resources, "The Uranium Path to the Bomb: 1942–1944," in *The Manhattan Project: An Interactive History,* www.osti.gov/opennet/manhattan-project-history/Events/1942-1944_ur/1942-1944_uranium.htm.
35. Rhodes, *Making of the Atomic Bomb,* 494–495; U.S. Atomic Energy Commission, *Manhattan District History,* book II, *Gaseous Diffusion (K-25) Project,* vol. 4, *Construction* (Washington, DC: U.S. Atomic Energy Commission, May 19, 1947), 3.28.

36. Renfrew, Science History Institute Interview; AEC, *Manhattan District History,* book VII, vol. 1, G.27–G.28; Christopher Bryson, *The Fluoride Deception* (New York: Seven Stories Press, 2004), 70.
37. Susan Freinkel, *Plastic: A Toxic Love Story* (Boston: Houghton Mifflin, 2011), 16.
38. Charles M. A. Stine, "Molders of Better Destiny," *Science* 96, no. 2492 (October 2, 1942): 305–311.
39. Groves would later describe these fears in a confidential letter to the Senate committee on atomic matters: "If the plaintiffs succeed in establishing that all or part of the damage alleged was caused by the DuPont plants operating for the Manhattan Project, DuPont and the government would be forced to pay the damages under its contract with the company. . . . I am keeping in close personal touch with the matter from day to day." L. R. Groves to Senator Brian McMahon, February 18, 1946, Groves Papers.
40. Philip Sadtler, transcript of interview, March 23, 1993, Bryson and Griffiths Papers.
41. Groves to McMahon, February 18, 1946.
42. Eileen Welsome, *The Plutonium Files: America's Secret Medical Experiments in the Cold War* (New York: Dial Press, 1999); U.S. Congress, House Committee on Energy and Commerce, *American Nuclear Guinea Pigs: Three Decades of Radiation Experiments on U.S. Citizens,* report from the Subcommittee on Energy Conservation and Power, (Washington, DC: U.S. Government Printing Office, 1986).
43. Col. Stafford Warren, "Purpose and Limitations of the Biological and Health Physics Research Program," memorandum to the file, July 30, 1945, Bryson and Griffiths Papers; Hymer L. Friedell, Lt. Col. Medical Corps, to Brigadier General K. D. Nichols, "Future Medical Research Program," memorandum, February 26, 1946, Bryson and Griffiths Papers.
44. Specifically, Conant's letter described the "extraordinary physiological properties" of a naturally occurring compound called fluoroacetate, but it cautioned that the findings might apply more broadly: "As an organic chemist I think I should point out [that] it is conceivable that similar effects would occur with any fluorinated organic acid." Many widely used synthetic fluorocarbons, including PFOA, fall into this category. James B. Conant to Dr. T. H. Wensel, Clinton Engineering Works, October 6, 1943, Bush-Conant Files.
45. Unattributed memorandum to Major H. L. Friedell, U.S. Engineer Office, Knoxville, TN, "Subject: Conference on Fluorine Metabolism 6 January 1944," January 15, 1944, Bryson and Griffiths Papers.
46. Renfrew, Science History Institute Interview; G. H. Gehrmann to Captain J. L. Ferry, E.I. du Pont de Nemours & Company Service Department, "Re: Fatalities Due to Exposure to HF in the Kinetic Chemicals Plant," May 5, 1944, Bryson and Griffiths Papers.
47. Capt. Ferry to Col. Warren, "Fatalities Occurring from a By-Product of T.F.E.," February 2, 1944, Bryson and Griffiths Papers.

48. District Engineer Ruhoff to Dr. H. T. Wensel, Clinton Engineer Works, March 30, 1944, Bryson and Griffiths Papers.

CHAPTER 5: A CATCH-22

1. Bea Peterson, "Many in the Hoosick Falls Community Pay Their Respects to John Ersel Hickey" *Eastwick Press* (Petersburgh, NY), March 8, 2013; Zeke Wright, "Ex-Hoosick Falls Trustee John 'Ersel' Hickey Dies," *Troy (NY) Record*, March 1, 2013.
2. Kieron Kramer, "Three Local Families Are Memorialized by the County," *Eastwick Press* (Petersburgh, NY), May 24, 2013.
3. "New York: Hoosick: Chemical Spill In River," *New York Times*, July 5, 2001.
4. C8 Science Panel, "Probable Links Report," October 29, 2012. (The panel's work also led to dozens of peer-reviewed, published papers, which are listed on the C8 Science Panel website: www.c8sciencepanel.org/publications.html.)
5. New York State, Senate Standing Committee on Health and Senate Standing Committee on Environmental Conservation, *Public Hearing: Drinking Water Contamination*, August 30, 2016 (written testimony of Mayor David Borge of Hoosick Falls).
6. Katie Eastman, "Record Voter Turnout in Hoosick Falls," *Spectrum News* (Albany, NY), March 16, 2016.
7. Matt Kelly, "Hoosick Falls Seeks a Downtown Vision," *Bennington (VT) Banner*, October 28, 1992.
8. Saint-Gobain, "Saint-Gobain in Figures," accessed in August 2024, www.saint-gobain.com/en/finance/saint-gobain-figures.
9. Caitlin Randall, "Outsourced Jobs and Poisoned Water: An American Town Fights for Survival," *Wilson Quarterly*, summer 2016; Bea Peterson, "Dougherty Corner to Be Developed at Last," *Eastwick Press* (Petersburgh, NY), July 25, 2014.
10. Evan Lawrence, "Diamond in the Rough? In Downtown Hoosick Falls, Projects Could Set Stage for Rebirth," *Hill Country Observer* (Cambridge, NY), September 15, 2014; Chris Mays, "Creativity Continues at HAYC3," *Bennington (VT) Banner*, July 14, 2015; "The Official Opening of Brown's New Walloomsac Brewery," *Eastwick Press* (Petersburgh, NY), October 3, 2014; Chris Churchill, "Churchill: Downtown Rises in Hoosick Falls," *Times Union* (Albany, NY), September 13, 2014; Leigh Hornbeck, "Builder Part of Effort to Revitalize Hoosick Falls," *Times Union* (Albany, NY), August 5, 2017.
11. Rich Elder to Jim Hurlburt, Village of Hoosick Falls, "PFOA Testing," August 14, 2014; Dave Borge to Rich Elder, Rensselaer County, "Fwd: Fwd: Preliminary Perfluorosurfactant Data," October 16, 2014; Brendan J. Lyons, "Emails Show Early Confusion Over Hoosick Falls Water Pollution," *Times Union* (Albany, NY), February 8, 2016.
12. Stephanie J. Frisbee et al. "The C8 Health Project: Design, Methods, and

Participants," *Environmental Health Perspectives* 117, no. 12 (December 2009): 1873–1882, doi:10.1289/ehp.0800379.

13. Borge confirmed that he had denied Michael access to the raw drinking water but said he had done so based on guidance from the state Department of Health.
14. Unlike her husband, Sue preferred Pepsi to tea or water. As a result, she had much lower PFOA levels than her friends and relatives who drank the water regularly or worked in the factories long term.
15. Eurofins Eaton Analytical, Inc., "Laboratory Report 326785," prepared for the Village of Hoosick Falls, October 16, 2014.
16. New York State Senate, *Drinking Water Contamination,* written testimony of Mayor David Borge; Ric DiDonato to Dave Borge, "Water," December 17, 2014; David Borge to Rich Elder, Rensselaer County, and Timothy E. Vickerson, New York State Department of Health, "Water Follow Up," January 18, 2015.
17. Dave Borge to Kimberly Evans McGee, New York State Department of Health, "Hoosick Falls Water Status," December 26, 2014.
18. Rich Elder, Rensselaer County Department of Health, to Kimberly Evans McGee and Timothy E. Vickerson, New York State Department of Health, "Fwd: Fwd: Report for Perfluorosurfactants Samples," October 17, 2014.
19. The description of Borge's meeting with Saint-Gobain is based on contemporaneous emails between him and county health officials. However, the company repeatedly asserted that it probably wasn't responsible for the contamination. Around the time of the initial meeting between Borge and company officials, Saint-Gobain prepared a standby press release and a Q&A, which included the following statement: "We don't believe that our plant is the source of the PFOA in the well, as we stopped using the dispersion material [that contains PFOA] more than 10 years ago." This claim was deeply misleading given PFOA's well-documented persistence in the environment. See Saint-Gobain Performance Plastics, "Village Water Quality Information," December 17, 2014, Baker v. Saint-Gobain Performance Plastics Corp., No. 17-3942 (2nd Cir. 2018), Plaintiff's Exhibit S130; David Borge to Rich Elder, Rensselaer Health Department, "Update," December 24, 2014.
20. The subsidiary, Furon Company, which Saint-Gobain acquired in 1999, had operated the McCaffery Street plant since 1996. By its own admission, Saint-Gobain continued emitting PFOA from the McCaffrey Street plant until at least 2003. See Charles Flemming, "Saint-Gobain Plans to Buy Furon in a Bid to Boost Plastics Division," *Wall Street Journal,* September 21, 1999; "Memorandum of Opinion and Order," Benefits Committee of Saint-Gobain v. Key Trust, No. 160 F. Supp. 2d 816 (N.D. Ohio 2001), February 5, 2001.
21. Julia DiCorleto, Saint-Gobain, to David Borge, "Meeting on January 9th at 10 am at Saint-Gobain McCaffrey Street Site," December 31, 2014; David Borge to Ken Holbrook et al., "Hoosick Falls Water Follow Up," January 23, 2015; David Borge to Julia DiCorleto, "Thanks!," January 9, 2015.

22. Saint-Gobain executives specifically suggested that Borge exclude other village officials from an early meeting for this reason. Julia Corletto, Saint-Gobain, to David Borge, "Re: Participants," January 18, 2015.
23. Edward J. Canning, Saint-Gobain, to David Borge, "Re: Follow Up from Our Meeting Last Thursday," January 28, 2015.
24. New York State Senate, *Drinking Water Contamination,* written testimony of Mayor David Borge.
25. Minutes, Village of Hoosick Falls Board Meeting, December 9, 2014.
26. Specifically, the letter stated that "carbon filters are known to be effective in screening out this element. The current filters are state-of-the-art poly filters which include carbon filtration."
27. Borge was interviewed for this book in 2017 and 2021, but he declined to participate further and did not respond to emails seeking comment about his December 2014 letter to village residents. David B. Borge, Mayor, Village of Hoosick Falls, to Hoosick Falls residents, December 20, 2014; Eurofins Eaton Analytical, "Laboratory Report 328480," prepared for the Village of Hoosick Falls, November 2014.
28. Minutes, Village of Hoosick Falls Board Meeting, December 9, 2014.
29. Arlene Anderson to Roger Sokol and Teresa Boepple-Swider, New York Department of Health, "PFOA in Hoosick Falls Water," December 19, 2014; Roger Sokol to Arlene Anderson, "FW: PFOA in Hoosick Falls Water," December 19, 2014; Arlene Anderson to Philip Sweeney, "Re: Follow up on the PFOA in Hoosick Falls Water," December 23, 2014.

CHAPTER 6: BIOLOGICAL DYNAMITE

1. "A-Bomb Suit Filed by Salem Peach Growers," *Courier-Post* (Camden, NJ), October 18, 1945; "W. Gotshalk, Ex-Official," *Courier-Post* (Camden, NJ), September 16, 1982.
2. Leslie R. Groves to Senator Brian McMahon, Chairman, Special Committee on Atomic Energy, February 18, 1946, Groves Papers; H. E. Stokinger, University of Rochester, to Colonel Stafford L. Warren, U.S. Engineer Office, Oak Ridge, TN, March 1, 1946, Bryson and Griffiths Papers; Philip Sadtler, transcript of interview, March 23, 1993, Bryson and Griffiths Papers.
3. Sadtler, interview, March 23, 1993. Subsequent research commissioned by Manhattan Project officials during this period confirmed Sadtler's findings. See Harold C. Hodge to Colonel Stafford L. Warren, U.S. Engineer Office, Manhattan District ex, Oak Ridge, TN, March 1, 1946, Bryson and Griffiths Papers.
4. Cooper B. Rhodes, Lt. Colonel, Infantry, "Subject: Peach Crop Cases (Kille et al. v. du Pont)," memorandum to file, May 2, 1946, Bryson and Griffiths Papers; H. E. Stokinger to Stafford L. Warren, U.S. Engineer Office, Oak Ridge, TN, March 1, 1946, Groves Papers; Sadtler, interview, March 23, 1993.

5. "A-Bomb Suit Filed."
6. F. G. Gosling, "The Manhattan Project: Making the Atomic Bomb," U.S. Department of Energy, Office of History and Heritage Resources, January 2010, 19–20.
7. The mortality figures for the bombings are based on a 1977 investigation conducted by a team of international scientists with support from the World Health Organization. By contrast, a study conducted immediately after the attacks by the U.S.-backed Joint Commission for the Investigation of the Atomic Bomb in Japan estimated that seventy thousand people died in Hiroshima and that forty thousand died in Nagasaki. For more information on these estimates and the reasons for the discrepancy, see Alex Wellerstein, "Counting the Dead at Hiroshima and Nagasaki," *Bulletin of the Atomic Scientists,* August 2020; Shoichirō Kawasaki and Yasuo Miyake, eds., *A Call from Hibakusha of Hiroshima and Nagasaki: Proceedings International Symposium on the Damage and After-Effects of the Atomic Bombing of Hiroshima and Nagasaki, July 21–August 9, 1977* (Tokyo: Japan National Preparatory Committee by Asahi Evening News, 1978).
8. "A-Bomb Suit Filed"; Maj. C. A. Tanney to William C. Gotshalk, U.S. Engineer Office, New York, NY, September 24, 1945, Bryson and Griffiths Papers.
9. Stokinger to Warren, March 1, 1946.
10. Lt. Col. Cooper B. Rhodes, "Kille et al. (12 separate cases) v. E. I. du Pont de Nemours & Co., Inc. et al (Peach Crop Cases)," memorandum to file, February 13, 1946, Bryson and Griffiths Papers; Stokinger to Warren, March 1, 1946.
11. Lt. Col. Cooper Rhodes to Major General Groves, "Subject: Proposed Conference with Mr. Willard B. Kille and Mr. A.J. Gorand," memorandum, March 7, 1946, Groves Papers; Lt. Col. Cooper Rhodes to General Nichols, "Subject: Conference with Mr. Willard B. Kille," memorandum, March 25, 1946, Bryson and Griffiths Papers.
12. Willard B. Kille to Maj. General L. R. Groves, March 26, 1946, Groves Papers.
13. Harold C. Hodge to Col. S. L. Warren, Medical Section, U.S. Engineer Office, Oak Ridge, TN, May 1, 1946, Bryson and Griffiths Papers.
14. Advisory Committee on Human Radiation Experiments, *Final Report* (Washington, DC: U.S. Government Printing Office, 1995); Eileen Welsome, *The Plutonium Files: America's Secret Medical Experiments in the Cold War* (New York: Dial Press, 1999).
15. Hodge to Warren, May 1, 1946.
16. Rhodes, "Subject: Peach Crop Cases (Kille et al. v. du Pont)," May 2, 1946; Harold C. Hodge to Colonel Stafford L. Warren, U.S. Engineer Office, Manhattan District ex, Oak Ridge, TN, February 27, 1946, Bryson and Griffiths Papers.
17. See, for example, Hodge to Warren, May 1, 1946; Bryson, *The Fluoride Deception,* 78–91; H. C. Hodge, "Fluoride Metabolism: Its Significance in Water Fluoridation," *Journal of the American Dental Association* 52, no. 3 (March 1956): 307–314, doi:10.14219/jada.archive.1956.0058.
18. Hodge was explicit about the purpose of the animal studies: "Rats, and later

perhaps rabbits and a dog or two, can be exposed to various levels of [hydrogen fluoride] . . . to find what level of HF inhaled air will produce blood levels such as those reported for humans residing in the areas." Hodge to Warren, March 1, 1946; Harold C. Hodge, "Research Plans for the Division of Pharmacology 1946–47," University of Rochester, Bryson and Griffiths Papers.

19. "Annual Report on the Observation of Fluorides," October 25, 1954, Kehoe Archives.
20. Hodge, "Research Plans for the Division of Pharmacology 1946–47"; Hodge to Warren, March 1, 1946.
21. U.S. Department of Energy, Staff of the Advisory Committee on Human Radiation Experiments, "Documents Retrieved from Oak Ridge Operations: The Atomic Energy Commission's Declassification Review of Reports on Human Experiments and the Public Relations and Legal Liability Consequence," memorandum to committee members, December 6, 1994.
22. Neil McKay, *A Chemical History of 3M, 1993-2000* (Minneapolis: 3M Chemical, Film & Allied Products Group, 1991), 2, 7, 24–25, State of Minnesota v. 3M Co., 845 N.W.2d 808 (Minn. 2014), Plaintiff's Exhibit 1365.
23. McKay, *A Chemical History of 3M,* 27, 158; Clifford E. Hicks, "Wanted: Jobs for a Trillion New Chemicals," *Popular Mechanics,* March 1952.
24. McKay, *A Chemical History of 3M,* 25–30.
25. McKay, *A Chemical History of 3M,* 44–45.
26. Hicks, "Jobs for a Trillion New Chemicals."
27. McKay, *A Chemical History of 3M,* 44–45.
28. "New Plastic Produced," *Pittsburgh Sun-Telegraph,* April 2, 1950.
29. John Butenhoff, Geary Olsen, and Andrea Pfahles-Hutchens, "PFOS Case Study," March 11, 2005, State of Minnesota v. 3M Co., Plaintiff's Exhibit 2696.
30. Michael Pollan, "What's Eating America," *Smithsonian Magazine,* July 2006.
31. Claudia Flavell-White, "Dermot Manning and Colleagues at ICI: Plastic Fantastic," Chemical Engineers Who Changed the World, *Chemical Engineer,* November 1, 2011; Susan Freinkel, *Plastic: A Toxic Love Story* (Boston: Houghton Mifflin, 2011), 96–108; "Ralph Wiley, Inventor of Saran," *Brinewell* (Michigan Division, Dow North America), fall/winter 1993, 13–16. Saran wrap was made of polyvinylidene chloride, or PVDC, which is similar, but not identical, to the plastic typically referred to as polyvinyl chloride, or PVC.
32. Jeffrey L. Meikle, *American Plastic: A Cultural History* (New Brunswick, NJ: Rutgers University Press, 1997), 148–151; "Women Risk Life and Limb in Bitter Battle for Nylons," *Augusta (GA) Chronicle,* December 16, 1945.
33. Freinkel, *Plastic: A Toxic Love Story,* 27.
34. The author personally examined these unpublished reports while conducting research in Kehoe Archives. For additional information, see Devra Davis, *The Secret History of the War on Cancer* (New York: Basic Books), xvi–xvii, 81–82; Markowitz and Rosner, *Deceit and Denial,* 174–175.
35. Robert A. Kehoe, "Review of *Silent Spring,* by Rachel Carson," prepared for *Farm Quarterly,* 1963, Kehoe Archives.

36. Wilhelm Hueper, "Adventures of a Physician in Occupational Cancer," Wilhelm C. Hueper Paper, History of Medicine Division, National Library of Medicine, (hereafter cited as Hueper Papers), 46–47, 138–142, 148; Christopher Sellers, "Discovering Environmental Cancer: Wilhelm Hueper, Post–World War II Epidemiology, and the Vanishing Clinician's Eye," *American Journal of Public Health* 87, no. 11 (November 1997): 1824–1835, doi:10.2105/ajph.87.11.1824; Gerald J. Fitzgerald, "Chemical Warfare and Medical Response During World War I," *American Journal of Public Health* 98, no. 4 (April 1998): 611–625, doi:10.2105/AJPH.2007.11930.
37. Davis, *War on Cancer,* 74–77; Hueper, "Adventures of a Physician in Occupational Cancer," 138–142, 148.
38. W. C. Hueper, *Occupational Tumors and Allied Diseases* (Springfield, IL: Charles C. Thomas, 1942), 668; Sellers, "Discovering Environmental Cancer."
39. David Michaels, "When Science Isn't Enough: Wilhelm Hueper, Robert A. M. Case, and the Limits of Scientific Evidence in Preventing Occupational Bladder Cancer," *International Journal of Occupational and Environmental Health* 1, no. 3 (1995): 278–288; H. L. Stewart, "Wilhelm Carl Heinrich Hueper, M.D.," *Toxic Pathology* 2, no. 1 (1974): 9–11.
40. Hueper, "Adventures of a Physician in Occupational Cancer," 193; W. C. Hueper, "Carcinogenic Studies on Water-Insoluble Polymers," *Pathologia et Microbiologia* 24, no. 1 (1961).
41. The censorship of Hueper's work was scrutinized by Congress and the media. See Davis, *War on Cancer,* 101–103; Hueper, "Adventures of a Physician in Occupational Cancer," 174, 177–182; "Was Report on Cancer Held Back? House Investigates Paper Suppression," *Springfield (MO) News-Leader,* April 10, 1958; "Some Industries Accused of Cancer Danger Whitewash," *Associated Press,* March 30, 1960.
42. See, for example, B. S. Oppenheimer et al., "Further Studies of Polymers as Carcinogenic Agents in Animals," *American Association for Cancer Research* 15, no. 5 (June 1955): 333–340.
43. Markowitz and Rosner, *Deceit and Denial,* 178–182; Sarah A. Vogel, *Is It Safe? BPA and the Struggle to Define the Safety of Chemicals* (Berkeley: University of California Press, 2012), 8, 16.
44. James Delaney, "Peril on Your Food Shelf," *American Magazine,* July 1951.
45. U.S. Congress, House Select Committee to Investigate the Use of Chemicals in Food Products, *Chemicals in Food Products,* 82nd Cong. (1951), 1353–1575; "Cancer Inducing Chemicals, Extension of Remarks of Hon. James J. Delaney of New York in the House of Representatives, Friday, August 22, 1958," U.S. Congressional Record, Proceedings and Debates of the 85th Cong., 2nd Sess., vol. 104, pt. 25 (1951), A7738–A7739.
46. Karen S. Miller, *The Voice of Business: Hill & Knowlton and Postwar Public Relations* (Chapel Hill: University of North Carolina Press, 2000), 128–129; "Minutes of the Thirteenth Meeting of the Directors of the Manufacturing Chemists' Association, Inc.," New York City, November 13, 1951, accessed via

Toxic Docs; "Minutes of the Seventy-Second Meeting of the Directors of the Manufacturing Chemists' Association, Inc.," New York City, September 10, 1957, accessed via Toxic Docs.

47. See, for example, "Role of Chemicals in Food Betterment: Report Tells of Sciences Contribution to More and Tastier Eating," *Joplin (MO) Globe,* February 14, 1954.
48. Vogel, *Is It Safe? BPA and the Struggle to Define the Safety of Chemicals,* 36.
49. Charles Vaughan, "French Inventor Making Jobs for Americans," *Indianapolis News,* December 7, 1961; Funderburg, "Making Teflon Stick."
50. Under normal conditions, fluoropolymers like Teflon aren't toxic because they don't interact with other substances the way most PFAS do. The concern among industry scientists was mostly about processing aids like PFOA and the chemicals that Teflon released as it broke down. M. S. Kessler, MD, DuPont Plant Physician, to Robert A. Kehoe, March 9, 1950, Kehoe Archives; Robert A. Kehoe to Harold D. Fields, Jr., February 5, 1958, Kehoe Archives.
51. "The Immediate Toxicity of the Thermal Decomposition Products of Teflon and Kel-F," April 26, 1954, Kehoe Archives.
52. Kehoe to Fields, February 5, 1958; Robert A. Kehoe to Harold D. Fields, Jr., January 28, 1958, Kehoe Archives.
53. Harland Manchester, "Amazing Fluorocarbons Promise Stainproof Clothing," *Popular Science,* January 1959.
54. The information about Henne's involvement in the Manhattan Project is based on archival bomb-project documents. See, for example, "Minutes of the Meeting of the Fluorocarbon Group," June 19, 1942, Miller Papers; Albert Henne to Robert A. Kehoe, October 15, 1958, Kehoe Archives.
55. Henne to Kehoe, October 15, 1958; Robert A. Kehoe to Albert L. Henne, October 31, 1958, Kehoe Archives.
56. Thomas G. Hardie to Robert A. Kehoe, March 23, 1959, Kehoe Archives.
57. Robert A. Kehoe to Henry P. McNulty, April 27, 1959, Kehoe Archives.
58. Funderburg, "Making Teflon Stick"; Joe Schwartz, "The Right Chemistry: Teflon-Coated 'Happy Pan' Brought Joy, Then Concerns," *Montreal Gazette,* May 9, 2015.
59. Dorothy Hood to Gerald A. Arenson, Polychemical Department, Research and Development Division, Experimental Station, November 9, 1964, United States Environmental Protection Agency v. E.I. du Pont de Nemours and Company, No. TSCA-HQ-2004-0016, Complainant's Exhibit 4.
60. J. E. Higginbotham to B. A. Herbert et al., "Teflon Scrap," memorandum, October 28, 1998, U.S. EPA Public Docket AR226-1445; P. S. Koepp and R. L. Lewis to M. J. Miller, "Investigation of Current Teflon Waste Disposal," memorandum, May 13, 1975, Plaintiffs' Third Motion for Summary Judgment, E.I. du Pont de Nemours & Co. C-8 Personal Injury Litigation, No. 2:13-md-2433 (S.D. Ohio 2015), Exhibit C, Tab 4; Deposition of Robert W. Rickard, August 27, 2014, E.I. du Pont de Nemours & Co. C-8 Personal Injury Litigation.

61. McKay, *Chemical History of 3M*, 157–159; Sharon Lerner, "The U.S. Military Is Spending Millions to Replace Toxic Firefighting Foam with Toxic Firefighting Foam," *The Intercept*, February 10, 2018.
62. Alan Deutschman, "The Fabric of Creativity," *Fast Company*, December 1, 2004; Alex Ward, "An All-Weather Idea," *New York Times*, November 10, 1985; W. L. Gore & Associates Official Company Timeline, accessed in October 2024, www.gore.com/about/the-gore-story/timeline.
63. Ward, "An All-Weather Idea."
64. James G. Chandler et al., "Polytetrafluoroethylene Large Vein Replacements and High-Altitude Treks: Footnotes and Footprints from Ben Eiseman's Panoply of Interests," *Journal of Vascular Surgery: Venous and Lymphatic Disorders* 1, no. 3 (July 2013): 320–323, doi: 10.1016/j.jvsv.2013.01.006.
65. G. L. Kelly and B. Eiseman, "Development of a New Vascular Prosthetic: Lessons Learned," *Archives of Surgery* 117, no. 10 (October 1982): 1367–1370, doi:10.1001/archsurg.1982.01380340081019.
66. Ward, "An All-Weather Idea"; Gore-Tex Brands, "Our History," www.gore-tex.com/about/history.
67. Ward, "An All-Weather Idea."

CHAPTER 7: BLOOD SECRETS

1. Rachel Carson, *Silent Spring* (New York: Houghton Mifflin Harcourt, 1962), 17, 20, 34–35, 101, 115–119.
2. Carson, *Silent Spring*, 7, 37, 69, 240–241, 244, 258–261.
3. Eliza Griswold, "How 'Silent Spring' Ignited the Environmental Movement," *New York Times Magazine*, September 2012; "Rachel Carson Dies of Cancer," *New York Times*, April 15, 1964; Meir Rinde, "Richard Nixon and the Rise of American Environmentalism," *Distillations Magazine* (Science History Institute), June 2017.
4. Gladwin Hill, "Congress Plans New Push to Control Toxic Chemical Products," *New York Times*, July 8, 1975; Laurence Richter et al., "Producing Ignorance Through Regulatory Structure: The Case of Per- and Polyfluoroalkyl Substances (PFAS)," *Sociological Perspectives* 60, no. 4 (October 2020): 631–656, doi.org/10.1177/073112142096482.
5. Gary Keith, *Eckhardt: There Once Was a Congressman from Texas* (Austin: University of Texas Press, 2007), 258–259; "How They Shaped the Toxic Substances Law," *Chemical Week*, April 27, 1977.
6. U.S. Environmental Protection Agency, "Procedures for Chemical Risk Evaluation Under the Amended Toxic Substances Control Act," 40 CFR Part 702, January 19, 2017.
7. John Donahue, "Historic Landscaping," *Cultural Resources Management, A National Park Service Technical Bulletin* 9, no. 2 (April 1986): 1–2.
8. Plunkett, Science History Institute Interviews; James F. Clarity and Warren Weaver, Jr., "To Teflon or Not to Teflon," *New York Times*, January 18, 1986.

9. R. J. Burger to Fluoropolymer Supervisors, "C-8 (FC-143) Status Report," memorandum, December 15, 1981, U.S. EPA Public Docket AR 226-1392.
10. According to a confidential memo summarizing her conversation with the plant physician, Younger Lovelace Power, Bailey also had other questions and concerns: "She would like to know if the 3M studies found any malformations other than right in the eye. She is especially concerned about the eye lid. She would also like to be able to read the reports from the . . . animal studies herself." J. F. Doughty, Washington Works, to R. D. Ingalls, Wilmington, "Questions on C8 Status Report," memorandum, December 18, 1981, U.S. EPA Public Docket AR 226-1393; Deposition of Bruce W. Karrh, MD, April 14, 2004, Jack W. Leach et al. v. E.I. du Pont de Nemours & Co., No. 01-C-698 (W.Va. Cir. Ct. 2002), 234–235, 257–272, (hereafter cited as Karrh Deposition).
11. G. H. Crawford, "Fluorocarbons in Human Blood Plasma," August 20, 1975, State of Minnesota v. 3M Co., Plaintiff's Exhibit 1118.
12. The methods used to identify molecules at the time were imprecise, and some 3M documents suggest the culprit was actually PFOS. In fact, both chemicals were accumulating in the bodies of people worldwide. J. W. Belisle, 3M, to H. E. Freier et al., "Fluorochemicals in Blood," memorandum, State of Minnesota v. 3M Co., Plaintiff's Exhibit 1142; Karrh Deposition, 29.
13. 3M Company, "Meeting Minutes—Meeting with H. C. Hodge," June 7, 1979, State of Minnesota v. 3M Co., Plaintiff's Exhibit 1210.
14. International Research and Development Corporation, "90-Day Subacute Rhesus Monkey Toxicity Study," report to 3M Company, January 2, 1979, State of Minnesota v. 3M Co., Plaintiff's Exhibit 1193; "Expert Report of Philippe Grandjean, MD, DMSc," prepared on behalf of Plaintiff State of Minnesota, State of Minnesota v. 3M Co., September 22, 2017, 49.
15. Robert A. Bilott, Taft Stettinius & Hollister LLP, to Christine T. Whitman, Administrator, U.S. Environmental Protection Agency, "Request for Immediate Governmental Action/Regulation Relating to DuPont's C-8 Releases in Wood County, West Virginia and Notice of Intent to Sue Under the Federal Clean Water Act, Toxic Substances Control Act, and Resource Conservation and Recovery Act," March 6, 2001 (hereafter cited as Bilott's March 2001 Letter), Exhibit 118.
16. 3M Company, "Meeting Minutes—Meeting with H. C. Hodge."
17. Riker Laboratories, "Oral Teratology Study of T-2998CoC in Rats," St. Paul, MN, January 1980, State of Minnesota v. 3M Co., Plaintiff's Exhibit 1252; George L. Hegg, Group Vice President, Chemicals, Film & Allied Products, 3M, to Document Control Officer, Chemical Information Division, Office of Toxic Substances (WH-557), U.S. EPA, March 20, 1981, U.S. EPA Public Docket AR226-1373.
18. J. W. Raines to R. L. Richards, "Ammonium Perfluorooctanoate (C8) Rangefinder Study," memorandum, April 1, 1981, U.S. EPA Public Docket AR226-1377; R. J. Burger to Supervision through Division Superintendents, "C-8 Communication," memorandum, March 31, 1981, U.S. EPA Public Docket AR226-1374.

19. "Pregnancy Outcome Questionnaire," U.S. EPA Public Docket AR-226-1384; Bruce W. Karrh, MD, to Carl De Martino, "Epidemiology Study—C-8 (FC-143)," memorandum, April 2, 1981, U.S. EPA Public Docket AR226-1379.
20. Robert Bilott to Richard H. Hefter, Chief, High Production Volume Branch, U.S. EPA, "TSCA Section 8(e) Reporting for PFOA," July 3, 2003; S. S. Stafford to Dr. Y. L. Power, "Analysis of Blood Samples for Perfluorooctanoate," memorandum, May 19, 1981, U.S. EPA Public Docket AR226-1387; J. H. Todd, "C-8 Program Status," May 26, 1981, U.S. EPA Public Docket AR226-1388; William E. Fayerweather, "Study of Pregnancy Outcome in Washington Works Employees: Research Proposal," April 13, 1981, U.S. EPA Public Docket AR226-1381.
21. DuPont, undated internal memorandum on the birth-defects research, U.S. EPA Public Docket AR226-1383; Bilott, "TSCA Section 8(e) Reporting for PFOA."
22. Riker Laboratories, "Oral Teratology Study of T-2998CoC in Rats," April 1981, State of Minnesota v. 3M Co., Plaintiff's Exhibit 1267.
23. B. C. McKusick to J. W. Raines, "Report of FC-143 Teratogenic Studies to EPA," memorandum, March 15, 1982, U.S. EPA Public Docket AR226-1396; "C-8 (FC-143) Employee Communication," March 1, 1982, U.S. EPA Public Docket AR226-1395.
24. "Expert Report of Philippe Grandjean."
25. S. R. Laas to J. F. Doughty, PPD Washington Works, "Perfluorooctanoate (C8) in Water," memorandum, June 25, 1984, U.S. EPA Public Docket AR226-1399; J. F. Doughty to J. A. Schmid, "Summary of C-8 in Water Sampling Program," memorandum, August 29, 1984, U.S. EPA Public Docket AR226-1399; J. F. Doughty to J. A. Schmid, "Update on C-8 in Water Samples," memorandum, July 14, 1984, U.S. EPA Public Docket AR226-1399.
26. R. J. Zipfel to C. A. Dykes and J. A. Schmid, "C8 Control Program," memorandum, July 7, 1987, Bilott's March 2001 Letter, Exhibit 22; Karrh Deposition, 32.
27. J. A. Schmid to T. M. Kemp and T. L. Schrenk, "C8 Meeting Summary 5/22/84—Wilmington," memorandum, May 23, 1984, Plaintiffs' Third Motion for Summary Judgment, E.I. du Pont de Nemours & Co. C-8 Personal Injury Litigation, Exhibit C, Tab 9.
28. A. E. Reiner, "Fate of Fluorochemicals in the Environment: Biodegradation Studies of Fluorocarbons III," July 19, 1978, State of Minnesota v. 3M Co., Plaintiff's Exhibit 1153.

CHAPTER 8: THE TIPPING POINT

1. Brendan J. Lyons, "PCB Fight, Then Trial," *Times Union* (Albany, NY), December 14, 2014.
2. Kathryn A. Crawford et al., "Waxing Activity as a Potential Source of Exposure to Per- and Polyfluoroalkyl Substances (PFAS) and Other Environmental Contaminants Among the US Ski and Snowboard Community," *Environmental Research* 215, no. 3 (December 2022), doi:10.1016/j.envres.2022.114335.

3. The description of the meeting is based on interviews with Engel, Hickey, and Martinez, as well as contemporaneous notes taken by one of Engel's colleagues. Alexandra Becker, associate, Nolan & Heller, LLP, "Notes from 1/6/2015 Meeting with Mr. Michael Hickey and Dr. Marcus Martinez," January 6, 2015.
4. On behalf of John Hickey, "Claim for Compensation in a Death Case," State of New York, Workers' Compensation Board, February 24, 2015.
5. The details of the settlement offer come from interviews with Michael Hickey and David Engel. When asked about the matter, a spokesperson for Saint-Gobain said the company was "not able to offer comment on legal activity."
6. This comparison is based on data from the National Cancer Institute, which estimates that carcinoid cancer affects four in every hundred thousand adults. J. Morel Symons et al., "Confirmed and Potential Carcinoid Tumor Cases in the Dupont Cancer Registry," DuPont Epidemiology Program, Haskell Global Centers for Health and Environmental Services, October 30, 2007.
7. See, for example, Anglina Kataria et al., "Association Between Perfluoroalkyl Acids and Kidney Function in a Cross-Sectional Study of Adolescents," *Environmental Health* 14, no. 89 (November 2015), doi:10.1186/s12940-015-0077-9; Shefali Sood et al., "Association Between Perfluoroalkyl Substance Exposure and Renal Function in Children with CKD Enrolled in H3Africa Kidney Disease Research Network," *Kidney International Reports* 4, no. 11 (August 2019): 1641–1645, doi:10.1016/j.ekir.2019.07.017.
8. David Lukas, Senior Process Manager, MRB Group, to Joyce Donohue, U.S. Environmental Protection Agency, "PFOA—Hoosick Falls, N.Y.," March 19, 2015.
9. Marcus Martinez to Bob Ryan, "Fwd: New PFOA Exposure Study," August 22, 2015.
10. David Engel to David Borge, "PFOA Saint Gobain," October 16, 2015; David Engel to Thomas Ulasewicz, "Hoosick Falls/Saint-Gobain PFOA Matter," November 16, 2015.
11. Engel to Borge, "PFOA Saint Gobain"; David Engel to Phil Guy, Business Manager, Saint-Gobain Performance Plastics Corp. et al., "Re: Notice of Intent to Sue Under Section 7002(a)(1)(B) of the Resource Conservation and Recovery Act, 42 U.S.C.§6972(a)(1)(B)," memorandum, October 19, 2015.
12. Robin Bravender, "Regional Boss Recycles Everything—from Orange Peels to Organs," *E&E News by Politico*, May 29, 2014.
13. C.T. Male Associates, "PFOA Sampling Results for Soil and Groundwater (August–October 2015), Saint Gobain Performance Plastics, McCaffrey Street Plant," November 20, 2015; New York State Department of Health Wadsworth Center, Report no. EHS1500048345-SR-1, July 24, 2015.
14. David Borge to Mark Surdam et al., "Hoosick Falls Water Update," October 18, 2015; David Borge to John Pattison, Attorney for the Village of Hoosick Falls, "Water Status Update," October 21, 2015.
15. The information about how the firm spent its time is derived from a detailed analysis of its invoices. David Borge to John Patterson, "Re: Saint-Gobain PFOA," October 27, 2015; FitzGerald Morris Baker Firth, Glens Falls, NY, to

Village of Hoosick Falls, Invoice No. 64301, November 22, 2016; FitzGerald Morris Baker Firth, Glens Falls, NY, Village of Hoosick Falls, Invoice No. 65089, April 6, 2017.

16. The description of Judith Enck's actions is based on interviews as well as contemporaneous emails and notes written by Enck and acquired by the author via public records requests. Judith Enck to Paul Simon et al., "Saint Gobain, PFOA," November 13, 2015; Judith Enck, "Phone Call with NYS and J. Enck," handwritten notes, November 2015; Judith Enck, "Conference Call w/NY DEC and NY DOH," handwritten notes, December 7, 2015; Judith Enck, "Phone Call w/EPA R2 Staff and Dave Engel, Michael Hickey (Citizen)," handwritten notes, December 11, 2015.
17. Judith Enck, Regional Administrator, U.S. Environmental Protection Agency, to David Borge, Mayor, Village of Hoosick Falls, November 25, 2015.
18. New York State Department of Health, "Perfluorooctanoic Acid (PFOA) in Drinking Water, Hoosick Falls, New York: Long Fact Sheet," December 2015.

CHAPTER 9: WELCOME TO BEAUTIFUL PARKERSBURG, WEST VIRGINIA

1. John G. Britvec to W. M. Stewart, September 25, 1995, Robert A. Bilott, Taft Stettinius & Hollister LLP, to Christine T. Whitman, Administrator, U.S. Environmental Protection Agency, "Request for Immediate Governmental Action/Regulation Relating to DuPont's C-8 Releases in Wood County, West Virginia and Notice of Intent to Sue Under the Federal Clean Water Act, Toxic Substances Control Act, and Resource Conservation and Recovery Act," March 6, 2001, Exhibit 57.
2. See the following sources, all of which were attached to Bilott's March 2001 Letter: "Dupont Washington Works Dry Run Landfill Chronology: 1995–1998," Exhibit 5; Laidley Eli McCoy, Director, West Virginia Division of Environmental Protection, to Mr. H. David Ramsey, DuPont, November 27, 1996, Exhibit 67; West Virginia Division of Environmental Protection, Order No. 3850, December 31, 1996, Exhibit 69.
3. "The Tennant Farm Health Herd Investigation: Cattle Team Report," December 23, 1999, U.S. EPA Public Docket AR226-1473.
4. For more information on the meeting, see Robert Bilott, *Exposure: Poisoned Water, Corporate Greed, and One Lawyer's Twenty-Year Battle Against DuPont* (New York: Atria, 2019), 17–19.
5. Theo Colborn, Dianne Dumanoski, and John Peterson Myers, *Our Stolen Future: Are We Threatening Our Fertility, Intelligence, and Survival? A Scientific Detective Story* (New York: Dutton, 1996), 11–12, 21, 23, 26, 154–155, 234.
6. Michael Lemonick, "Heroes of the Environment: Theo Colborn," *Time,* October 17, 2007.
7. Holger Breithaupt, "A Cause Without a Disease," *European Molecular Biology Organization* 5, no. 1 (January 2004): 16–18; M. Y. Gross-Sorokin, S. D.

Roast, and G. C. Brighty, "Causes and Consequences of Feminisation of Male Fish in English Rivers," UK Environment Agency, science report SC030285/SR, July 2004.

8. E Carlsen et al., "Evidence for Decreasing Quality of Semen During Past 50 Years," *British Medical Journal* 305, no. 6854 (September 1992): 609–613, doi:10.1136/bmj.305.6854.609; Lawrence Wright, "Silent Sperm," *New Yorker,* January 15, 1996; Hagai Levine et al., "Temporal Trends in Sperm Count: A Systematic Review and Meta-Regression Analysis," *Human Reproduction Update* 23, no. 6 (November 2017): 646–659, doi:10.1093/humupd/dmx022.
9. Wright, "Silent Sperm."
10. Susan C. Nagel et al., "Relative Binding Affinity—Serum Modified Access (RBA-SMA) Assay Predicts the Relative in Vivo Bioactivity of the Xenoestrogens Bisphenol A and Octylphenol," *Environmental Health Perspectives* 105, no. 1 (1997): 70–76; F. S. vom Saal et al., "A Physiologically Based Approach to the Study of Bisphenol A and Other Estrogenic Chemicals on the Size of Reproductive Organs, Daily Sperm Production, and Behavior," *Toxicology and Industrial Health* 14, nos. 1–2 (1998): 239–260, doi:10.1177/074823379801400115.
11. See, for example, Laura N. Vandenberg et al., "Hormones and Endocrine-Disrupting Chemicals: Low-Dose Effects and Nonmonotonic Dose Responses," *Endocrine Reviews* 33, no. 3 (August 2012): 378–455, doi:10.1210/er.2011-1050; Jennifer T. Wolstenholme et al., "Gestational Exposure to Bisphenol A Produces Transgenerational Changes in Behaviors and Gene Expression," *Endocrinology* 153, no. 8 (August 2012): 3828–3838, doi:10.1210/en.2012-1195.
12. U.S. Environmental Protection Agency, "Endocrine Disruptor Screening Program," *Federal Register* 63, no. 154 (August 11, 1998): 42852–42855.
13. Didier Bourguignon, "The Precautionary Principle," European Parliamentary Research Service, December 2015.
14. Gerald L. Kennedy, Jr., et al., "The Toxicology of Perfluorooctanoate," Baker v. Saint-Gobain Performance Plastics Corp., Plaintiff's Exhibit S34.
15. Riker Laboratories, Inc./3M Company, "Two Year Oral (Diet) Toxicity/Carcinogenicity Study of Fluorochemical FC-143 in Rats," August 29, 1987, State of Minnesota v. 3M Co., Plaintiff's Exhibit 1337; John L. Butenhoff et al., "Genotoxicity, Carcinogenicity, Developmental Effects and Reproductive Effects of Perfluorooctanoate: A Perspective from Available Animal and Human Studies," prepared for the Association of Plastics Manufacturers of Europe and the Society of Plastics Industry, December 19, 2002; Lisa B. Biegel et al., "Mechanisms of Extrahepatic Tumor Induction by Peroxisome Proliferators in Male CD Rats," *Toxicological Sciences* 60, no. 1 (March 2001): 44–55, doi:10.1093/toxsci/60.1.44.
16. R. Perkins to A. Sethre, "Ammonium Perfluorooctanoate TSCA 8(e) Submission," memorandum, October 5, 1989, State of Minnesota v. 3M Co., Plaintiff's Exhibit 1360.
17. Judy Walrath, Epidemiology Section, to Y. L. Power, MD, Washington Works, Polymers, "Washington Works: Surveillance Data Mortality and Cancer Incidence," memorandum with attached report, August 28, 1992, U.S. EPA Public Docket AR226-1546.

18. Frank D. Gilliland, "Fluorocarbons and Human Health: Studies in an Occupational Cohort" October 1992, U.S. EPA Public Docket AR226-0473; Frank D. Gilliland and Jack S. Mandel, "Mortality Among Employees of a Perfluorooctanoic Acid Production Plant," *Journal of Occupational Medicine* 35, no. 9 (September 1993): 950–954, doi:10.1097/00043764-199309000-00020.
19. See, for example, 3M's involvement with the Advancement of Sound Science Coalition and the American Council on Science and Health. Elisa K. Ong and Stanton A. Glantz, "Constructing 'Sound Science' and 'Good Epidemiology': Tobacco, Lawyers, and Public Relations Firms," *American Journal of Public Health* 91, no. 11 (November 2001): 1749–1757, doi:10.2105/ajph.91.11.1749; Andy Kroll and Jeremy Schulman, "Leaked Documents Reveal the Secret Finances of a Pro-Industry Science Group," *Mother Jones,* October 28, 2013.
20. Sharon Boone to J. P. Glas et al., "C-8 Ammonium Perfluorooctanoate Fluorosurfactant Strategies and Plans," memorandum, September 28, 1994, attachment to Gerald Kennedy Deposition, vol. 1, United States Environmental Protection Agency v. E.I. du Pont de Nemours and Company, No. TSCA-HQ-2004-0016, Complainant's Exhibit 115 (hereafter cited as Kennedy Deposition).
21. DuPont, "C-8 Project Phase I Review," January 27, 1997, Baker v. Saint-Gobain Performance Plastics Corp., Plaintiff's Exhibit S48; DuPont, "C-8 Project Phase IIc Review," July 30, 1998, Plaintiff's Exhibit S52; John M. Migliore, "C-8 BIODEGRADATION," email to Rodger Zipfel, September 14, 1999, Baker v. Saint-Gobain Performance Plastics Corp, Plaintiff's, Exhibit S11.
22. These companies' motives were spelled out in the minutes from an April 1997 meeting of fluorochemical manufacturers in Hattersheim am Main, Germany: "A key reason for proceeding with further studies . . . was to address both classification and labelling concerns and also food contact applications, both of which were governed by proscriptive regulation in the EU." See "Notes of a Meeting of an Ad-Hoc Group of Toxicology Representatives of APFO Producers and Users to Discuss Forward Plans for Toxicological Research on APFO," Hoechst Toxicology Laboratory, Hattersheim, April 18, 1997, attachment to Kennedy Deposition; "C-8 Monkey Study," presentation to the Association of Plastic Manufacturers in Europe, attachment to Kennedy Deposition.
23. Robert F. Pinchot to Richard J. Angiullo et al., "Results from the C-8 Monkey Study," attachment to Kennedy Deposition, October 22, 1999; Covance Laboratories Inc., "Final Report: 26-Week Capsule Toxicity Study with Ammonium Perfluorooctanoate in Cynomolgus Monkeys Report," prepared for APME Ad-Hoc APFO Toxicology Working Group, December 18, 2001, State of Minnesota v. 3M Co., Plaintiff's Exhibit 1812; "Notes of a Meeting of an Ad-Hoc Group of Toxicology Representatives of APFO Producers and Users"; R. J. Zipfel et al., "C-8 Ammonium Perfluorooctanoate Fluorosurfactant: Strategies and Plans," September 15, 1994, attachment to Kennedy Deposition.
24. Robert Pinchot, Mike McCord, and Trini Garza, "C8 Strategy Review," March 3, 2000, Baker v. Saint-Gobain Performance Plastics Corp., Plaintiff's Exhibit S59.
25. For more on the American Red Cross study, see Geary W. Olsen et al., "Per-

fluorooctanesulfonate and Other Fluorochemicals in the Serum of American Red Cross Adult Blood Donors," *Environmental Health Perspectives* 111, no. 16 (December 2003): 1892–1901, doi:10.1289/ehp.6316. For more on the archived blood-sample research, see 3M, "Fluorochemicals in Human Blood," draft memorandum, 1998, State of Minnesota v. 3M Co., Plaintiff's Exhibit 1479; 3M, "Working Memorandum on Data Quality Assessment," 1998, U.S. EPA Public Docket AR226-0036.

26. Joseph Weber, "3M's Big Cleanup," *Businessweek,* June 5, 2000.
27. John Butenhoff et al., "Genotoxicity, Carcinogenicity, Developmental Effects and Reproductive Effects of Perfluorooctanoate: A Perspective from Available Animal and Human Studies," December 19, 2002, U.S. EPA Public Docket AR226-1726.
28. Rich Purdy, "Occurrence of Perfluorooctane Sulfonate (PFOS) in Wildlife Part 1, Eagles and Albatrosses," October 16, 1998, State of Minnesota v. 3M Co., Plaintiff's Exhibit 1523.
29. Richard E. Purdy to Georjean L. Adams, "Risk to the Environment Due to the Presence of PFOS," email with attachment ("Pioneer Food Chain Risk Assessment of PFOS"), December 3, 1998, State of Minnesota v. 3M Co., Plaintiff's Exhibit 1533; Richard E. Purdy to 3M, March 28, 1999, State of Minnesota v. 3M Co., Plaintiff's Exhibit 1001.
30. Weber, "3M's Big Cleanup."
31. 3M Company, "3M FC Issue Communications Plan," May 24, 1999, State of Minnesota v. 3M Co., Plaintiff's Exhibit 1587; John L. Butenhoff, "Toxicological Research Program in Perfluorinated Chemistries," slide presentation, March 29, 2008, State of Minnesota v. 3M Co., Plaintiff's Exhibit 2206.
32. G. W. Olsen et al., "An Epidemiologic Investigation of Reproductive Hormones in Men with Occupational Exposure to Perfluorooctanoic Acid," *Journal of Occupational and Environmental Medicine* 40, no. 7 (July 1998): 614–622, doi:10.1097/00043764-199807000-00006.
33. Gilliland, "Fluorocarbons and Human Health," 225; "Expert Report of Philippe Grandjean, MD, DMSc," prepared on behalf of Plaintiff State of Minnesota, State of Minnesota v. 3M Co., September 22, 2017, 49.
34. Thomas J. DiPasquale to Georjean L. Adams, "Re: Re: 8e Follow Up—Fish," March 26, 1999, State of Minnesota v. 3M Co., Plaintiff's Exhibit 1003.
35. Rich Purdy to 3M, "Resignation," March 28, 1999, State of Minnesota v. 3M Co., Plaintiff's Exhibit 1001.
36. Richard E. Purdy to Eric A. Reiner et al., "Rich Purdy's Resignation," March 28, 1999, State of Minnesota v. 3M Co., Plaintiff's Exhibit 1002; Charles Auer to Dan C. Hakes, "Follow-Up on Perfluorochemicals Request," April 11, 2000, State of Minnesota v. 3M Co., Plaintiff's Exhibit 1665.
37. Charles Auer to Barbra Leczynski et al., "Phaseout of PFOS," May 16, 2000, U.S. EPA Public Docket AR226-0629; 3M Company, "3M Phasing Out Chemical In Scotchgard®," July 1, 2000.
38. David Barboza, "E.P.A. Says It Pressed 3M for Action on Scotchgard Chemical," *New York Times,* May 19, 2000.

39. Vaughn Hagerty, "Legacy of C8 Persists in the Cape Fear Region," *Wilmington Star-News* (NC), August 19, 2017; Expert Report of Steven Amter, E.I. du Pont de Nemours & Co. C-8 Personal Injury Litigation (hereafter cited as Expert Report of Steven Amter), 70.
40. Bilott, *Exposure*, 61–64.
41. Bernard J. Reilly to unidentified recipient, "Aug 13," August 13, 2000, United States Environmental Protection Agency v. E.I. du Pont de Nemours and Company, No. TSCA-HQ-2004-0016, Complainant's Exhibit 104; Deposition of Bernard J. Reilly, July 2, 2015, E.I. du Pont de Nemours & Co. C-8 Personal Injury Litigation.
42. Bilott, *Exposure*, 76.
43. The provisional standard, recommended by DuPont scientists in 1988 and formally adopted in 1991, was set in parts per billion. (One part per billion is equal to 1,000 parts per trillion.) Anthony J. Playtis to H. David Ramsey, "C-8 Exposure Limits," memorandum, August 4, 2000, Bilott's March 2001 Letter, Exhibit 3; J. A. Schmid to J. F. Doughty, "Update on C-8 in Water Samples," memorandum, June 14, 1984, Bilott's March 2001 Letter, Exhibit 18; "Answer and Request for Hearing," August 11, 2004, United States Environmental Protection Agency v. E.I. du Pont de Nemours and Company.
44. "Dry Run Landfill: C-8 History Summary," U.S. EPA Public Docket AR226-1476.
45. A. C. Huston to H. V. Bradley, "Elimination of Supernate Ponds," June 25, 1986, E.I. du Pont de Nemours & Co. C-8 Personal Injury Litigation, Plaintiffs' Third Motion for Summary Judgment, Exhibit C, Tab 15; A. C. Huston to H. V. Bradley, "Purchase of Lubeck Public Service District Property," July 16, 1987, E.I. du Pont de Nemours & Co. C-8 Personal Injury Litigation, Plaintiffs' Third Motion for Summary Judgment, Exhibit C, Tab 17.
46. Bilott, *Exposure*, 81.
47. Bernard J. Reilly to unknown recipient, "May 7," May 7, 2001, U.S. EPA v. E.I. du Pont de Nemours and Company, Complainant's Exhibit 104.
48. "DuPont Washington Works Dry Run Landfill Chronology, 1995–1998"; Thomas R. Waldron to Ronald W. Meloon, "Dry Run Landfill," memorandum, October 7, 1996, Bilott's March 2001 letter, Exhibit 5.
49. "Attorney Client/Work Product Privilege: Win for DuPont," E.I. du Pont de Nemours & Co. C-8 Personal Injury Litigation, Plaintiffs' Third Motion for Summary Judgment, Exhibit C, Tab 30.
50. Bilott, *Exposure*, 88.
51. The 972-page document is cited throughout the endnotes as Bilott's March 2001 Letter.
52. Reilly, the DuPont lawyer, wrote candidly about judge's refusal to grant the gag order in an email to his son: "Court yesterday did not agree to shut up plaintiff lawyer in our Parkersburg situation and today he testifies [at the] EPA hearing and will try to slam us one more time. A miracle the press has not picked this up yet. . . . Oh well, I told the clients to settle many moons ago. Too bad they are still in denial and don't think things can get worse, wrong

again." Bernard J. Reilly to Bernie Reilly, "Mar 27," March 27, 2001, U.S. EPA v. E.I. du Pont de Nemours and Company, Complainant's Exhibit 104. See also Transcript of Proceedings, March 29, 2001, Tennant v. E.I. du Pont de Nemours & Company, Inc, No. 6:99-0488 (S.D. W.Va. 1998); Expert Report of Steven Amter, 65–66.

53. James M. Cox, Manager, Lubeck Public Service District, to Lubeck Public Service District Customers, October 31, 2000; Deposition of Craig Skaggs, March 6, 2002, Jack W. Leach et al. v. E.I. du Pont de Nemours and Company.

CHAPTER 10: A ROCK IN THE MACHINE

1. Bower v. Westinghouse Elec. Corp., 206 W. Va. 133 (W. Va. 1999); William S. Rogers and Florice E. Engler, "A Growing Number of States Recognize Medical Monitoring Claims," *Bloomberg Law,* April 22, 2010.
2. Robert Griffin Interrogatory, August 13, 2014, The Little Hocking Water Assn., Inc., v. E.I. du Pont de Nemours and Company, 90 F. Supp. 3d 746 (S.D. Ohio 2015), Plaintiff's Exhibit 43.
3. Expert Report of Steven Amter, 68–70.
4. Bernard J. Reilly to Bernie Reilly, "Oct 13," October 13, 2001, United States Environmental Protection Agency v. E.I. du Pont de Nemours and Company, No. TSCA-HQ-2004-0016, Complainant's Exhibit 104; Bernard J. Reilly to Bernie Reilly, "Oct 12," October 12, 2001, United States Environmental Protection Agency v. E.I. du Pont de Nemours and Company, Complainant's Exhibit 104.
5. Reilly to Reilly, "Oct 13."
6. Rosalind Adams and Lisa Song, "One-Stop Science Shop Has Become a Favorite of Industry—And Texas," *Inside Climate News/Center for Public Integrity,* December 2014; David Andrews and Melanie Benesh, "Mr. Pay to Spray, Michael Dourson," Environmental Working Group, September 2017; Toxicology Excellence for Risk Assessment, "Annual Funding Sources," accessed in May 2024, www.tera.org/about/FundingSources.html.
7. For examples, see Richard Denison, "Proof in Pudding: EPA Toxics Nominee Dourson Has Consistently Recommended 'Safe' Levels for Chemicals That Would Weaken Health Protections," Environmental Defense Fund, September 22, 2017.
8. Timothy S. Bingham to Robert W. Rickard, Director, DuPont Haskell Laboratory for Health and Environmental Sciences et al., "Prospective Contractors for PFOA Criteria Review," August 21, 2000, E.I. du Pont de Nemours & Co. C-8 Personal Injury Litigation, Exhibit W.
9. The West Virginia Department of Environmental Protection (WVDEP) maintains that TERA's selection was unrelated to DuPont's appeals since it outsourced the selection process to the National Institute for Chemical Studies, an organization backed by regional business leaders. Besides industry representatives, the team TERA assembled included three officials from the EPA and one

from the Agency for Toxic Disease Registry. WVDEP, "Final Ammonium Perfluorooctanoate (C8) Assessment of Toxicity Team (CATT) Report," August 2002.

10. Bernard J. Reilly to Bernie Reilly, "Nov 28," November 28, 2001, U.S. EPA v. E.I. du Pont de Nemours and Company, Complainant's Exhibit 104.
11. WVDEP, "Final Ammonium Perfluorooctanoate (C8) Assessment."
12. Robert Bilott, *Exposure: Poisoned Water, Corporate Greed, and One Lawyer's Twenty-Year Battle Against DuPont* (New York: Atria, 2019), 167–169.
13. In an interview, Dee Ann Staats, the DuPont-funded WVDEP consultant who ordered the documents destroyed, said she was following departmental rules on document retention; the department maintained it had no such agencywide policy.
14. "Injunction Order Directed to Dee Ann Staats, Ph.D., and the West Virginia Department of Environmental Protection," June 26, 2002, Jack W. Leach et al. v. E.I. du Pont de Nemours and Company; Bilott, *Exposure,* 164, 173–174.
15. Ken Ward, Jr., "DuPont Lawyer Edited DEP's C8 Media Releases," *Charleston (WV) Sunday Gazette-Mail,* July 3, 2005.
16. Ken Ward, Jr., "Group Calls for Timmermeyer to Quit over C8," *Charleston (WV) Gazette-Mail,* July 7, 2005.
17. U.S. Environmental Protection Agency, "DuPont Agrees to Lower Limit of PFOA in Drinking Water," news release, March 12, 2009; Robert A. Bilott to Susan Hedman, Shawn Garvin, and Randy Huffman, "In the Matter of: E.I. du Pont de Nemours and Company," January 20, 2015.
18. U.S. Environmental Protection Agency, "Perfluorooctanoic Acid, Fluorinated Telomers; Request for Comment, Solicitation of Interested Parties for Enforceable Consent Agreement Development, and Notice of Public Meeting," *Federal Register* 68, no. 73 (April 16, 2003): 18626–18633.
19. Notably, more than 96 percent of the samples tested contained quantifiable levels of PFAS. In some product categories (nonstick cookware, for example), the concentrations were modest. In other categories, like dental floss, the levels varied widely by brand. The product types that were most likely to contain large quantities of PFAS were carpeting, floor wax, and stain-resistant clothing and upholstery. Zhishi Guo et al., "Perfluorocarboxylic Acid Content in 116 Articles of Commerce," U.S. Environmental Protection Agency, National Risk Management Research Laboratory, March 2009; Charles M. Auer, Director, Office of Pollution Prevention and Toxics, U.S. Environmental Protection Agency, to Oscar Hernandez, Mary Ellen Weber, and Ward Penberthy, "Revision of PFOA Hazard Assessment and Next Steps," memorandum, September 27, 2002.
20. Mark J. Strynar and Andrew B Lindstrom. "Perfluorinated Compounds in House Dust from Ohio and North Carolina, USA," *Environmental Science & Technology* 42, no. 10 (April 2008): 3751–3756, doi:10.1021/es7032058.
21. All information related to DuPont's Memorandum of Understanding can be found in U.S. EPA Public Docket OPPT-2004-0113; information related to 3M's Memorandum of Understanding can be found in EPA Public Docket OPPT-2004-0112.

22. Robert Bilott, remarks at the Taconic Mountain Student Water Conference, Bennington College, Bennington, VT, May 20, 2017.
23. The group was formally known as the Society of the Plastics Industry's Fluoropolymers Processors Group. See, for example, Allen C. Weidman, Executive Director, Society of the Plastics Industry, "Processors Organizational Meeting," October 31, 2003, Burdick v. Tonoga Inc., 191 A.D.3d 1220 (N.Y. App. Div., 2021), Plaintiff's Exhibit 34; Don Duncan, Society of the Plastics Industry, to Tom McCarthy, Taconic Plastics Inc., "Re: Request for Participation in Dispersion Processors Material Study," October 17, 2003, Burdick v. Tonoga Inc, Plaintiff's Exhibit 30; Carol Goodermote to Andy Kawczak, "C8 environmental issues," May 31, 2002, Burdick v. Tonoga, Plaintiff's Exhibit 20.
24. Barr Engineering Company and KHA Consulting, LLC., "Final Report: Dispersion Processor Material Balance Project," sponsored by the Society of the Plastics Industry's Fluoropolymers Manufacturers Group, February 2005.
25. Saint-Gobain, "Project 'Tymor': Fluorosurfactant Team," Baker v. Saint-Gobain Performance Plastics Corp., Plaintiff's Exhibit S122; Saint-Gobain, "PFOA," presentation to R. Caliari, CREE, September 22, 2004, Baker v. Saint-Gobain Performance Plastics Corp., Plaintiff's Exhibit S126; Deposition of Ed Canning, Baker v. Saint-Gobain Performance Plastics Corp., Plaintiff's Exhibit S105, p. 276.
26. Ruth A. Jamke to Susan Lindsey, "Info for Key Customers and VF meeting on 4/26," February 22, 2006, Baker v. Saint-Gobain Performance Plastics Corp., Plaintiff's Exhibit S124.
27. Brian Ross, "Can Nonstick Make You Sick?," *20/20*, ABC, November 13, 2003.
28. John R. Bowman to Thomas L. Sager and Martha L. Rees, "Lubeck-Dawn-Jackson Note," November 9, 2000, E.I. du Pont de Nemours & Co. C-8 Personal Injury Litigation, Plaintiffs' Third Motion for Summary Judgment, Exhibit C, Tab 35.
29. U.S. EPA v. E.I. du Pont de Nemours and Company, No. TSCA-HQ-2004-0016; Mark Glassman, "EPA Says It Will Fine DuPont For Holding Back Test Results," *New York Times*, July 9, 2004; Ken Ward, Jr., "DuPont Faces Criminal Probe," *Charleston (WV) Gazette*, May 20, 2005.
30. This figure is based on the decline in DuPont's stock prices between July 2004 and July 2005.
31. David Michaels, *Doubt Is Their Product: How Industry's Assault on Science Threatens Your Health* (Oxford: Oxford University Press, 2008), 86; Elisa K. Ong, "Tobacco Industry Efforts Subverting International Agency for Research on Cancer's Second-Hand Smoke Study," *Lancet* 355, no. 9211 (April 2000): 1253–1259, doi:10.1016/S0140-6736(00)02098-5; Paul D. Thacker, "The Weinberg Proposal," *Environmental Science & Technology* 40, no. 9 (June 2006): 2868–2869.
32. P. Terrence Gaffney, Weinberg Group, to Jane Brooks, Vice President, Special Initiatives DuPont de Nemours & Company, "Perfluorooctanoic Acid (PFOA)," April 29, 2003.
33. See, for example, H. Edward Dunkelberger III, Senior Director, Product

Defense, Weinberg Group, to Emma Burton, Esq., Crowell & Moring LLP, invoice cover letter, October 8, 2003; P. Terrence Gaffney to Scott L. Winkelman, Esq., Crowell & Moring LLP, invoice cover letter, November 11, 2003; P. Terrene Gaffney to Scott L. Winkelman, December 15, 2003; H. Edward Dunkelberger to Richard C. Bingham, Director, Technology Planning, E.I. du Pont de Nemours and Company, letter with attached report, "Preliminary Expert Candidates," September 29, 2003. (The preceding documents were entered into the public records as attachments to the deposition of H. Edward Dunkelberger, E.I. du Pont de Nemours & Co. C-8 Personal Injury Litigation.)

34. U.S. Environmental Protection Agency, "President Clinton Names W. Michael McCabe as Deputy Administrator at the Environmental Protection Agency," press release, November 16, 1999.
35. Deposition of W. Michael McCabe, December 7, 2007, E.I. du Pont de Nemours & Company v. Misty Scott and all others similarly situated (hereafter cited as McCabe Deposition), 95–96.
36. McCabe Deposition, 126–130.
37. Stephen T. Washburn et al., "Exposure Assessment and Risk Characterization for Perfluorooctanoate in Selected Consumer Articles," *Environmental Science & Technology* 39, no. 11 (June 2005): 3904–3910, doi:10.1021/es048353b.
38. Gilbert L. Ross, M.D., "Teflon and Human Health: Do the Charges Stick?," American Council on Science and Health, March 18, 2005.
39. Marian Burros, "As Teflon Troubles Pile Up, DuPont Responds with Ads," *New York Times,* February 8, 2006; Ken Ward, Jr., "Dupont Distorted C8 Study: Scientists," *Charleston (WV) Sunday Gazette-Mail,* October 14, 2007.
40. Elizabeth Whelan, "Teflon Concerns Don't Stick," *Washington Times,* July 6, 2005.
41. The wording was apparently meant to convey that the agency was unaware of evidence linking the levels of exposure among the *general public* to health problems. While this may have been technically accurate—the health data available at the time came mostly from studies on animals and workers—it was nonetheless deeply misleading, given the growing evidence that PFOA was a human carcinogen. See Stephen L. Johnson, EPA Administrator, to Charles O. Holliday, Jr., Chairman and Chief Executive Officer, DuPont, January 25, 2006.
42. U.S. Environmental Protection Agency, Science Advisory Board, "SAB Review of the EPA's Draft Risk Assessment of Potential Human Health Effects Associated with PFOA and Its Salts," May 30, 2006.
43. Susan M. Stalnecker to Michael McCabe, "Urgent: Script," February 16, 2006, E.I. du Pont de Nemours & Co. C-8 Personal Injury Litigation, Exhibit McCabe 45.
44. McCabe Deposition, 175–179; Deposition of Kathleen Forte, E.I. du Pont de Nemours & Co. C-8 Personal Injury Litigation, November 12, 2014, 260–261.
45. U.S. Environmental Protection Agency, "100 Percent Participation and Commitment in EPA's PFOA Stewardship Program," news release, March 2, 2006.
46. Michael Janofsky, "DuPont to Pay $16.5 Million for Unreported Risks," *New York Times,* December 15, 2015.

47. Class Action Settlement Agreement, Leach v. E.I. du Pont de Nemours and Company, November 17, 2004.
48. Stephanie J. Frisbee et al., "The C8 Health Project: Design, Methods, and Participants," *Environmental Health Perspectives* 117, no. 12 (December 2009): 1873–1882, doi:10.1289/ehp.0800379.
49. Frisbee et al., "C8 Health Project."

CHAPTER 11: "THEY POISONED THE WORLD"

1. Robert Bilott, *Exposure: Poisoned Water, Corporate Greed, and One Lawyer's Twenty-Year Battle Against DuPont* (New York: Atria, 2019), 24; C8 Science Panel, "Probable Link Reports," October 29, 2012; Heyong-Moo Shin et al., "Retrospective Exposure Estimation and Predicted Versus Observed Serum Perfluorooctanoic Acid Concentrations for Participants in the C8 Health Project," *Environmental Health Perspectives* 119, no. 12 (December 2011): 1760–1765, doi:10.1289/ehp.1103729; Robert Bilott, remarks at the Taconic Mountain Student Water Conference, Bennington College, Bennington, VT, May 20, 2017.
2. **On ringed seals:** Rossana Bossi, Frank F. Riget, and Rune Dietz, "Temporal and Spatial Trends of Perfluorinated Compounds in Ringed Seal (*Phoca hispida*) from Greenland," *Environmental Science & Technology* 39, no. 19 (September 2005): 7416–7422, doi:10.1021/es0508469. **On ducks:** Simon Sharp et al., "Per- and Polyfluoroalkyl Substances in Ducks and the Relationship with Concentrations in Water, Sediment, and Soil," *Environmental Toxicology and Chemistry* 40, no. 3 (July 2020): 846–858, doi:10.1002/etc.4818. **On dolphins:** Natalia Quinete et al., "Specific Profiles of Perfluorinated Compounds in Surface and Drinking Waters and Accumulation in Mussels, Fish, and Dolphins from Southeastern Brazil," *Chemosphere* 77, no. 6 (October 2009): 863–969, doi:10.1016/j.chemosphere.2009.07.079. **On indoor dust:** Tao Zhang et al., "Perfluorochemicals in Meat, Eggs and Indoor Dust in China: Assessment of Sources and Pathways of Human Exposure to Perfluorochemicals," *Environmental Science & Technology* 44, no. 9 (May 2010): 3572–3579, doi:10.1021/es1000159. **On apples:** Centre Analytical Laboratories, Inc., "Analysis of PFOS, FOSA and PFOA from Various Food Matrices Using HPLC, Electrospray/Mass Spectrometry," sponsored by Susan A. Beach, 3M, June 21, 2001, U.S. EPA Public Docket AR226-1030a. **On breast milk:** Judy S. LaKind et al., "Per- and Polyfluoroalkyl Substances (PFAS) in Breast Milk and Infant Formula: A Global Issue," *Environmental Research* 219, art. 115042 (February 2023), doi:10.1016/j.envres.2022.115042.
3. "USW Raises Concerns over Potentially Dangerous Chemical Leaking into Water Supply from DuPont Plant," *Environmental Protection,* June 1, 2005; John Reid Blackwell, "Chemical Found in River Near DuPont," *Richmond (VA) Times-Dispatch,* November 4, 2006; Stefan Baumgarten, "Steelworkers Claim PFOA Contamination at DuPont Site in Ohio," *Independent Commodity Intelligence Services,* August 12, 2005.

4. U.S. Securities and Exchange Commission, DuPont Shareholders for Fair Value on Behalf of E.I. du Pont de Nemours & Co., Notice of Exempt Solicitation, Delaware, April 12, 2007 (hereafter cited as Notice of Exempt Solicitation, 2007); U.S. Securities and Exchange Commission, DuPont Shareholders for Fair Value on Behalf of E.I. du Pont de Nemours & Co., Schedule 14A, Delaware, 2006 (hereafter cited as Schedule 14A, 2006).
5. Notice of Exempt Solicitation, 2007.
6. Gloria B. Post et al., "Occurrence and Potential Significance of Perfluorooctanoic Acid (PFOA) Detected in New Jersey Public Drinking Water Systems," *Environmental Science & Technology* 43, no. 12 (June 2009): 4547–4554, doi:10.1021/es900301s.
7. The description of the meeting is based on interviews with NJDEP staffers and deposition testimony from McCabe. See McCabe Deposition, 192–195, 234–249.
8. New Jersey Department of Environmental Conservation, "Contaminants of Emerging Concern," https://dep.nj.gov/srp/emerging-contaminants/.
9. Edward Anthony Emmett et al., "Community Exposure to Perfluorooctanoate: Relationships Between Serum Concentrations and Exposure Sources," *Journal of Occupational and Environmental Medicine* 48, no. 8 (August 2006): 759–770, doi:10.1097/01.jom.0000232486.07658.74.
10. McCabe Deposition, 244–245, 236–249.
11. Murphy gave the author a copy of the email from NJDEP's commissioner directing Murphy to withdraw the study. Lisa P. Jackson to Eileen Murphy, "Fwd: Draft PFOA Technical Paper," October 23, 2008.
12. Drinking Water Quality Institute, Meeting Minutes, New Jersey Environmental Infrastructure Trust Building, Princeton Pike, Lawrenceville, NJ, September 10, 2010.
13. Geoff Mulvihill, "Christie Restarts Water Board Left Dormant 3 Years," *Associated Press,* March 20, 2014; New Jersey Department of Environmental Protection, "Final Report of the Contaminants of Emerging Concern Work Group of the Science Advisory Board," September 2012.
14. New Jersey Department of Environmental Protection, "Administrative Order No. 2010-05," May 13, 2010.
15. U.S. Environmental Protection Agency, "Provisional Health Advisories for Perfluorooctanoic Acid (PFOA) and Perfluorooctane Sulfonate (PFOS)," January 8, 2009.
16. "PFOA Probe Ends Without Charges," *Chemical & Engineering News* 86, no. 43 (October 22, 2007).
17. Bea Peterson, "State of the Art Water Plant Is Worth It," *Eastwick Press* (Petersburgh, NY), August 28, 2009.
18. Bilott, *Exposure,* 280–290.
19. See, for example, Robert A. Bilott to Lisa P. Jackson, Director USEPA Headquarters et al., "Notice of Intent to Sue DuPont Under Citizen Suit Provisions of RCRA for PFOA Contamination of City of Parkersburg, West Virginia's Drinking Water Supply," February 23, 2009.

20. Ken Ward, Jr., "C8 Panel Moving Closer to Key Decisions," *Charleston (WV) Gazette,* July 31, 2011.
21. Ken Ward, Jr., "Wood Judge Blasts C8 Science Panel," *Charleston (WV) Gazette,* May 19, 2011.
22. C8 Science Panel, "Probable Link Evaluation of Pregnancy Induced Hypertension and Preeclampsia," December 5, 2011; C8 Science Panel, "Probable Link Reports."
23. Bilott, remarks at the Taconic Mountain Student Water Conference.
24. U.S. Environmental Protection Agency, "Third Unregulated Contaminant Monitoring Rule," May 2, 2012, www.epa.gov/dwucmr/third-unregulated-contaminant-monitoring-rule.
25. Marc C. Reisch, "First PFOA Lawsuit Against DuPont, Chemours Goes to Trial," *Chemical & Engineering News* 93, no. 37 (September 18, 2015): 6.
26. U.S. Securities and Exchange Commission, "The Chemours Company, Form 10-K, for the Fiscal Year Ended December 31, 2015."
27. Tiffany Kary and Jack Kaskey, "Can DuPont Spin Off Its Liabilities? Its 37 Active Chemical Plants—and EPA Problems—Will Go to Chemours," *Bloomberg,* July 2, 2015; Verified First Amended Complaint, The Chemours Company v. DowDuPont, et al., No. 2019-0351-SG (Del. 2020).
28. Alexander H. Tullo, "C&EN's Global Top 50 Chemical Companies of 2018," *Chemical & Engineering News* 97, no. 30 (July 29, 2019); Alexander H. Tullo, "DowDuPont Completes Final Split to Form DuPont and Corteva," *Engineering News* 97, no. 23 (June 7, 2019).
29. Complaint, Pinellas Marine Salvage, Inc., and John Mavrogiannis v. Kenneth R. Feinberg and Feinberg Rozen, LLP, and Gulf Coast Claims Facility, no. 2:11-cv-01987, February 25, 2011; Complaint, Selmer M. Salvesen v. Kenneth R. Feinberg, Feinberg Rozen, LLP, Gulf Coast Claims Facility, and William G. Green, Jr., no. 2:11-cv-02533, October 12, 2011; Jane Meinhardt, "Pinellas Marine Salvage Sues Feinberg over Oil Spill Claim," *Tampa Bay (FL) Business Journal,* March 11, 2011.

CHAPTER 12: THE RECKONING

1. Mariah Blake, "Welcome to Beautiful Parkersburg, West Virginia," *HuffPost Highline,* August 27, 2015; Sharon Lerner, "The Teflon Toxin," *The Intercept,* August 11, 2015.
2. Saint-Gobain's role in concealing information about PFOA hadn't yet come to light, so the company's actions didn't factor into Michael's outrage.
3. Judith Enck, Regional Administrator, U.S. Environmental Protection Agency, to David Borge, Mayor, Village of Hoosick Falls, November 25, 2015.
4. New York State Department of Health, "Perfluorooctanoic Acid (PFOA) in Drinking Water, Hoosick Falls, New York: Long Fact Sheet," December 2015.
5. Edward J. Canning, Saint-Gobain Performance Plastics, to David Borge, "Helpful Thoughts from Our Communications Folks," December 2, 2015.

6. Brendan J. Lyons, "A Danger That Lurks Below," *Times Union* (Albany, NY), December 14, 2015.
7. David Borge to Hoosick Falls Water Customers, December 18, 2015.
8. Daniel Bethencourt, "Crowd Calls for Snyder's Arrest at His Ann Arbor Home," *Detroit Free Press,* January 18, 2016; Heather Walker, "Protest over Flint Water Targets Gov. Snyder," *24 News Hour 8,* WOOD-TV (Grand Rapids, MI), January 14, 2016.
9. Governor Andrew M. Cuomo, "01 27 16 Albany Hoosick Falls Water," YouTube video, January 27, 2016, https://youtube.com/watch?v=3EdIwjeovUc; Hubert Wiggins, "Hoosick Falls Mayor Encouraged by Meeting with State Officials," *CBS-6 News Albany,* WRGB, January 27, 2016.
10. New York Governor's Press Office, "Governor Cuomo Announces Immediate State Action Plan to Address Contamination in Hoosick Falls," news release, January 27, 2016.
11. David Borge to Steve Mann, Deputy State Director, Office of Governor Andrew Cuomo, et al., "Hoosick Falls," January 24, 2016; Edward Damon, "Banks Will Lend When Hoosick Falls Water Safe to Drink," *Bennington (VT) Banner,* February 3, 2016.
12. Edward Damon, "Erin Brockovich to Hoosick Falls, N.Y. Residents: 'Everyone Entitled to Safe, Pure Water,'" *Berkshire (MA) Eagle,* January 31, 2016; CAT-TV Bennington, "Erin Brockovich Forum on Hoosick Falls Water Contamination," YouTube video, February 3, 2016, https://youtube.com/watch?v=AcGpvFTi60g.
13. Despite the law barring no-injury claims, during the 1980s and '90s several New York courts had awarded medical monitoring to plaintiffs who'd been exposed to hazardous substances based on the logic that exposure itself was a form of injury. However, the landscape had begun to shift in 2011, when the state's highest court rejected a medical-monitoring suit brought by a group of former smokers against Phillip Morris. In contrast to previous rulings, the judges refused to recognize medical monitoring as a cause of action absent an alleged physical injury. This led many experts to conclude that any future medical-monitoring claims were bound to be rejected, too.
14. Technically, the plants were run by one of Honeywell's predecessor companies, Allied Signal, which bought the company then known as Honeywell Inc. in 1999 and adopted the latter's name and brand identity. Class action complaint, Michael Hickey v. Saint-Gobain Performance Plastics Corp. and Honeywell International Inc., U.S. District Court: Northern District of New York, April 6, 2016.

CHAPTER 13: CLOUD NINE

1. Dick Valentinetti, Director, Air Pollution Control Division, Vermont Department of Environmental Conservation, to Charles Tilgner, Vice President and

Director, U.S. Operations and Engineering, ChemFab Corporation, "Re: Notice of Alleged Violation, Air Pollution Control Act," November 24, 1998; Chris Jones, Air Pollution Compliance Chief, Vermont Department of Environmental Conservation, to Charles Tilgner, "Enforcement Meeting on December 9, 1998," memorandum, December 8, 1998; Mitch Wertlieb and Kathleen Masterson," Saint-Gobain Rep Weighs In on PFOA Contamination," *Vermont Public Radio*, April 20, 2016.

2. Vermont Department of Environmental Conservation, "PFOA Summary for Legislators: PFOA Contamination in North Bennington," March 25, 2016; Jim Therrien, "Residents, Attorneys React to PFOA Suit Settlement," *Bennington (VT) Banner*, April 19, 2022; "Untangling Why Saint-Gobain Chose New Hampshire," *New Hampshire Public Radio*, April 22, 2016.
3. Mike Polhamus, "Teflon Town: Part 2," *VTDigger*, August 29, 2017.
4. New Hampshire Department of Environmental Services, "Saint-Gobain Site Investigation History," New Hampshire PFAS Response, 2023, www.pfas.des.nh.gov/pfas-occurrences/saint-gobain-performance-plastics/site-investigation-history; Michael Metcalf, Underwood Engineers Senior Project Manager, "MVD's PFAS Journey: Chapter One," Merrimack Village District Water Works, April 16, 2021.
5. New York State Department of Health, "NYS DOH Confirms PFOA in Petersburgh Water Supply at Levels Just Below Recent EPA Advisory," news release, February 21, 2016; New Hampshire Department of Environmental Services, "Southern New Hampshire PFOA Investigation: Public Meeting in Merrimack," June 29, 2016; Kenneth C. Crowe II, "Rensselaer County Tests Wells for PFOA," *Times Union* (Albany, NY), March 4, 2016; Timothy E. Vickerson, New York State Department of Health, to David Borge, "Re: Public Water System Sampling Results," February 26, 2015.
6. Paula Pierozan et al., "Perfluorooctanoic Acid (PFOA) Exposure Promotes Proliferation, Migration and Invasion Potential in Human Breast Epithelial Cells," *Archives of Toxicology* 92, no. 5 (May 2018): 1729–1739, doi.org/10.1007%2Fs00204-018-2181-4.
7. See, for example, Margaret Wood Hassan, Governor of New Hampshire, Andrew Cuomo, Governor of New York, and Peter Shumlin, Governor of Vermont, to Gina McCarthy, Administrator, U.S. Environmental Protection Agency, "Re: Public Water System Sampling Results," March 10, 2016.
8. Kenneth C. Crowe II, "Gibson Sends Letter to House Committee Seeking Hoosick Falls Probe," *Times Union* (Albany, NY), May 11, 2016; Scott Waldman, "More Prominent Democrats Call for Hoosick Falls Hearings," *Politico*, June 21, 2016.
9. U.S. Environmental Protection Agency, "Fact Sheet: PFOA & PFOS Drinking Water Health Advisories," May 19, 2016.
10. U.S. Environmental Protection Agency, "National Primary Drinking Water Regulations: Public Notification Rule," *Federal Register* 65, no. 87 (May 4, 2000): 25982; Nathaniel Rich, "The Story Behind the E.P.A.'s Contaminated

Water Revolution," *New York Times Magazine*, May 27, 2016; Bill Walker and David Andrews, "Drinking Water for 5.2 Million People Tainted by Unsafe Levels of PFCs," Environmental Working Group, May 23, 2016.

11. Technically, only residents of Hoosick Falls and the adjoining town of Hoosick were eligible for blood testing at the time. The sole reason Emily was able to qualify is that she had lived in the village previously.
12. C8 Science Panel, "C8 Study Results," memorandum, October 15, 2008, U.S. EPA Public Docket AR226-3835.
13. It is not unusual for the concentrations of PFOA in people's bodies to be much higher than in their drinking water. Studies show that residents of affected communities have blood levels that are one hundred times higher on average than the levels detected in their water supply, a sign of how rapidly the chemical builds up. See Gloria B. Post et al., "Perfluorooctanoic Acid (PFOA), an Emerging Drinking Water Contaminant: A Critical Review of Recent Literature," *Environmental Research* 116 (May 2012): 93–117, doi:10.1016/j.envres.2012.03.007.
14. Kyle Steenland et al., "Epidemiologic Evidence on the Health Effects of Perfluorooctanoic Acid (PFOA)," *Environmental Health Perspectives* 118, no. 8 (April 2010): 1100–1108, doi:10.1289/ehp.0901827.

CHAPTER 14: DIRTY WATER, DIRTY DEAL

1. "Hoosick Falls Residents Speak Out on Water Crisis," *New York Now*, season 2016, episode 27, aired on the New York Capital District's PBS affiliate, WMHT-TV, June 16, 2016.
2. For Loreeen Hackett's Twitter/X account, see PfoaProject NY (@pfoaprojectny1).
3. Scott Waldman, "Hoosick Falls Students Press Cuomo for New Water System," *Politico*, February 22, 2016.
4. Jesse McKinley, "Hoosick Falls Residents Take Anger over Tainted Water to New York's Capitol," *New York Times*, June 15, 2016.
5. The reporter posted footage to social media. See Ayla Ferrone (ABC News 10), "Several People from HF Storm Out as Hickey Says Hearings Aren't Needed," Facebook, July 8, 2016.
6. Brendan J. Lyons, "Congress to Probe Hoosick Falls Water Crisis," *Times Union* (Albany, NY), July 8, 2016; Jesse McKinley, "New York Senate to Hold Hearings on Hoosick Falls's Tainted Water," *New York Times*, July 8, 2016.
7. New York State, Senate Standing Committee on Health and Senate Standing Committee on Environmental Conservation, *Public Hearing: Drinking Water Contamination*, August 30, 2016, (statement of Michael Hickey, resident of Hoosick Falls).
8. New York State, *Drinking Water Contamination*, August 30, 2016, (statement of Rob Allen, resident of Hoosick Falls).
9. New York State Department of Health, "New York State Announces Water

from Village of Hoosick Falls Municipal Water System Is Safe to Drink," March 30, 2016.

10. New York State, Senate Standing Committee on Health, Senate Standing Committee on Environmental Conservation, Assembly Standing Committee on Environmental Conservation, and Assembly Standing Committee on Health, *Public Hearing: Water Quality and Contamination*, September 12, 2016 (written statement of Honeywell International Inc.).
11. Village of Hoosick Falls, "The Village of Hoosick Falls Agreement with Saint-Gobain and Honeywell: A Summary and Frequently Asked Questions of the Current Agreement," January 24, 2017; Marie J. French, "Hoosick Falls Residents Angered by Proposed Settlement with Saint-Gobain, Honeywell," *Politico*, January 12, 2017.
12. Village of Hoosick Falls, "Partial Settlement and Release Agreement," draft, released February 21, 2017.
13. Rob Allen, "Explanation of the PFOA Agreement: Hoosick Falls," February 26, 2017, YouTube video, https://youtube.com/watch?v=jiwrAM90K9o.
14. David Borge to Rich Elder et al., "Hoosick Falls Water Update," October 21, 2015; FitzGerald Morris Baker Firth, Glens Falls, NY, to Village of Hoosick Falls, Invoice No. 6430, November 22, 2016, and Invoice No. 65089, April 6, 2017.
15. The lawyer who authored the email said it was intended as "a joke amongst colleagues at a particularly frustrating time" in a process that had been difficult for everyone involved. John D. Aspland to Joan Gerhardt, Behan Communications, copying David Borge, "Re: Lyons on Agreement," January 27, 2017.
16. Behan Communications, Inc., to Mr. John D. Aspland, Esq., FitzGerald Morris Baker Firth, Invoices #280-4667-0116, #280-4667-0216, #280-4667-0316, #280-4667-0416, etc., January 12, 2016, through February 28, 2017; Behan Communications, "Case Study: More than Two Decades as GE's Communications Partner on Major Hudson River Cleanup," accessed in May 2024, www.behancommunications.com/case-studies; Behan Communications, "Representative Clients," accessed in September 2024, www.behancommunications.com/representative-clients.
17. Village of Hoosick Falls, "Special Meeting of the Board of Trustees," meeting minutes, June 13, 2017.
18. Wendy Liberatore, "Hoosick Falls Board Votes to Proceed with PFOA Lawsuits," *Times-Union* (Albany, NY), January 9, 2018.

CHAPTER 15: ACCIDENTAL ACTIVISTS

1. See, for example, Amber L. Cathey et al., "Exploratory Profiles of Phenols, Parabens, and Per- and Poly-Fluoroalkyl Substances Among NHANES Study Participants in Association with Previous Cancer Diagnoses," *Journal of Exposure Science & Environmental Epidemiology* 33 no. 5 (September 2023): 687–698, doi.org/10.1038/s41370-023-00601-6.

2. U.S. Environmental Protection Agency, "Statistics for the TSCA CBI Review Program," last updated April 8, 2024, www.epa.gov/tsca-cbi/statistics-tsca-cbi-review-program.
3. Catherine Clabby, "Local Scientists Uncovered Cape Fear River GenX Story," *North Carolina Health News,* October 18, 2017; Adam Wagner and Tim Buckland, "Chemours: GenX in River Since 1980," *Wilmington (NC) Star-News,* June 15, 2017.
4. Mei Sun et al., "Legacy and Emerging Perfluoroalkyl Substances Are Important Drinking Water Contaminants in the Cape Fear River Watershed of North Carolina," *Environmental Science and Technology Letters* 3, no. 12 (November 2016): 415–419, doi:10.1021/acs.estlett.6b00398.
5. Jian Zhou et al., "Three-Dimensional Spatial Distribution of Legacy and Novel Poly/Perfluoroalkyl Substances in the Tibetan Plateau Soil: Implications for Transport and Sources," *Environment International* 158, art. 107007 (January 2022), doi:10.1016/j.envint.2021.107007; Melanie R. Wells et al., "Per- and Polyfluoroalkyl Substances (PFAS) in Little Penguins and Associations with Urbanisation and Health Parameters," *Science of the Total Environment* 912, art. 169084 (February 2024), doi:10.1016/j.scitotenvs.2023.
6. Specifically, the EPA found that the tolerable daily dose for GenX was much lower than the one it had tabulated for PFOA (three nanograms per kilogram of body weight versus twenty nanograms per kilogram). The figure for PFOA has since been revised downward. For the EPA assessments, see U.S. Environmental Protection Agency, Office of Water, Health and Ecological Criteria Division, "Human Health Toxicity Values for Hexafluoropropylene Oxide (HFPO) Dimer Acid and Its Ammonium Salt, Also Known as GenX Chemicals," no. 822R-21-010, October 2021; U.S. Environmental Protection Agency, Office of Water, Health and Ecological Criteria Division, "Health Effects Support Document for Perfluorooctanoic Acid (PFOA)," no. 822-R-16-003, May 2016, 3–36. For the Swedish study, see Melissa Ines Gomis et al., "Comparing the Toxic Potency in Vivo of Long-Chain Perfluoroalkyl Acids and Fluorinated Alternatives," *Environment International* 113 (April 2018): 1–9, doi.org/10.1016/j.envint.2018.01.011.
7. See, for example, Stephan Brendel et al., "Short-Chain Perfluoroalkyl Acids: Environmental Concerns and a Regulatory Strategy Under REACH," *Environmental Sciences Europe* 30, no. 1, art 9 (February 2018), doi:10.1186/s12302-018-0134-4.
8. See, for example, Nadine Belkouteb et al., "Removal of Per- and Polyfluoroalkyl Substances (PFASs) in a Full-Scale Drinking Water Treatment Plant: Long-Term Performance of Granular Activated Carbon (GAC) and Influence of Flow-Rate," *Water Research* 182, art. 115913 (September 2020), doi:10.1016/j.watres.2020.115913.
9. Vaughn Hagerty, "Toxin Taints CFPUA Drinking Water," *Wilmington (NC) Star-News,* June 7, 2017.
10. Specifically, the groundwater contained up to 100,000 parts per trillion PFOA. The findings presented at the meeting were also recorded in a report submitted

to the New York Department of Environmental Conservation: C.T. Male Associates, "Data Summary Submittal (Through 2017), Saint-Gobain Performance Plastics, McCaffrey Street Site," May 7, 2018.

11. New York Department of Environmental Conservation, "State Superfund Site Classification Notice," Saint-Gobain Liberty Street, Site No. 442048, July 2017; NYDEC, "State Superfund Site Classification Notice," Former Oak Materials, John Street, Site No. 442049, July 2017; NYDEC, "State Superfund Site Classification Notice," Hoosick Falls Landfill, Site No. 442007, April 2019; U.S. Environmental Protection Agency, Office of Land and Emergency Management, "EPA Adds Superfund Sites to National Priorities List to Clean Up Contamination," July 31, 2017.
12. The description of the meeting is based on interviews with village officials and contemporaneous correspondence between Rob Allen and Tom Kinisky. Robert J. Allen, Mayor, Village of Hoosick Falls, to Thomas Kinisky, President and CEO, Saint-Gobain North America, August 22, 2017; Thomas Kinisky to Robert J. Allen, August 28, 2017.
13. The chemical in question, perfluorobutanoic acid, or PFBA, has been linked to liver damage, thyroid disease, reproductive problems, and delayed fetal development, among other ailments. A Danish study published at the height of the Covid-19 pandemic found that people with elevated PFBA levels were far more likely than the average person to develop severe symptoms and to die from the disease. **On the presence of PFBA in the village's water supply:** Lloyd R. Wilson, Director, Bureau of Water Supply Protection, New York State Department of Health, to the Honorable Rob Allen, Village of Hoosick Falls, February 12, 2018. **On PFBA's links to severe COVID-19:** Philippe Grandjean et al., "Severity of COVID-19 at Elevated Exposure to Perfluorinated Alkylates," *PloS One* 15, no. 12 (December 2020), doi:10.1371/journal.pone.0244815.
14. The figures for Wilmington are based on the number of chemicals Chemours releases into the part of the Cape Fear River that supplies the community's drinking water. **On Wilmington, NC:** Chemours Company, Fayetteville Works, "PFAS Non-Targeted Analysis and Methods Interim Report: Process and Non-Process Wastewater and Stormwater," June 30, 2020. **On Merrimack, NH:** Timothy H. Watkins, Director, EPA National Exposure Research Laboratory, to Clark Freise, Assistant Commissioner, New Hampshire Department of Environmental Services, "EPA PFAS Environmental Contamination Associated with Manufacturing Sites in New Hampshire, Laboratory Data Report #6: Non-Targeted PFAS Measurements in MM5 Sample Trains," letter with enclosed report, June 20, 2019. **On U.S. military bases:** Krista A. Barzen-Hanson et al., "Discovery of 40 Classes of Per- and Polyfluoroalkyl Substances in Historical Aqueous Film-Forming Foams (AFFFs) and AFFF-Impacted Groundwater," *Environmental Science and Technology* 51, no. 4 (January 2017): 2047–2057, doi:10.1021/acs.est.6b05843.
15. Nanyang Yu et al., "Nontarget Discovery of Per- and Polyfluoroalkyl Substances in Atmospheric Particulate Matter and Gaseous Phase Using Cryogenic Air Sampler," *Environmental Science and Technology* 54, no. 6 (March

2020): 3103–3113, doi.org/10.1021/acs.est.9b05457; Xiaowen Xia et al., "Nontarget Identification of Novel Per- and Polyfluoroalkyl Substances in Cord Blood Samples," *Environmental Science & Technology* 56, no. 23 (December 2022): 17061–17069, doi:10.1021/acs.est.2c04820.

16. Notably, the law was approved unanimously in the Senate and by a vote of 403–12 in the House. Frank R. Lautenberg Chemical Safety for the 21st Century Act, 15 U.S.C. § 2601 (2016), www.congress.gov/bill/114th-congress/house-bill/2576/all-actions.
17. Jessica A. Knoblauch, "Seven Ways This New Chemical Law Will Make You Safer," *Earthjustice,* October 24, 2016; Swati Meshram and Frank Ramos, "Understanding the 2016 Reform of the Toxic Substances Control Act (T.S.C.A)," *Save the Water,* May 2, 2017.
18. Lauren Coleman-Lochner, "Chemical Defender Put in Charge of EPA Unit Overseeing Toxins," *Bloomberg,* May 24, 2017; Eric Lipton, "Why Has the E.P.A. Shifted on Toxic Chemicals? An Industry Insider Helps Call the Shots," *New York Times,* October 21, 2017; Annie Snider and Alex Guillén, "EPA Staffers, Trump Official Clashed over New Chemical Rules," *Politico,* June 22, 2017.
19. Maria Hegstad, "Internal Memos Show EPA Offices' Fears over Narrowed Trump TSCA Rule," *Inside EPA's Risk Policy Report* 24, no. 43 (October 31, 2017): 1–11; Lipton, "Why Has the E.P.A. Shifted?"
20. U.S. Congress, Senate Committee on Environment and Public Works, *Hearing on the Nominations of Kristine Svinicki, Annie Caputo and David Wright to Be Members of the U.S. Nuclear Regulatory Commission, and the Nomination of Susan Bodine to Be Assistant Administrator of the Office of Enforcement and Compliance Assurance of the U.S. Environmental Protection Agency,* 115th Cong. (2017) (testimony of Susan Bodine); Sharon Lerner, "Donald Trump's Pick for EPA Enforcement Office Was a Lobbyist for Superfund Polluters," *The Intercept*, May 24, 2017.
21. EPA Press Office, "Widespread Praise for Dr. Michael Dourson," news release, July 17, 2017; Sheila Kaplan and Eric Lipton, "Chemical Industry Ally Faces Critics in Bid for Top E.P.A. Post," *New York Times,* September 17, 2017.
22. William Samuel, Director, Government Affairs Department, AFL-CIO, to John Barrasso, Chairman, Senate Committee on Environment and Public Works, and Thomas Carper, Ranking Member, Committee on Environment and Public Works, October 16, 2017; David Michaels, PhD, MPH, Professor of Environmental and Occupational Health, George Washington University, et al., to John Barrasso, Chairman, Committee on Environment and Public Works, and Tom Carper, Ranking Member, Committee on Environment and Public Works, October 17, 2017; Center for Environmental Health, "Over 100 Organizations Across the US Oppose Michael Dourson for the EPA," open letter signed by Center for Environmental Health and 108 other organizations, October 3, 2017.
23. U.S. Congress, Senate Committee on Environment and Public Works, *Hearing on the Nominations of Michael Dourson, Matthew Leopold, David Ross, and*

William Wehrum to Be Assistant Administrators of the Environmental Protection Agency, and Jeffery Baran to Be a Member of the Nuclear Regulatory Commission, 115th Cong. (2017), 10.

24. U.S. Congress, Senate Committee on Environment and Public Works, *Hearing on the Nominations of Michael Dourson,* 122–123.
25. Office of U.S. Senator Joe Manchin, "Manchin Statement on EPA Nominee Dourson," news release, October 25, 2017; Brian Murphy, "US Sens. Burr, Tillis Help Sink Trump's Pick for EPA Chemical Office," *News & Observer* (Raleigh, NC), December 13, 2017; Timothy Cama, "Collins 'Leaning Against' Trump EPA Chemical Nominee," *The Hill,* November 16, 2017.
26. Notably, Trump's plan was widely criticized by environmental groups for perpetuating the EPA's record of slow-walking PFAS regulation. For more on the EPA's actions, see U.S. Environmental Protection Agency, "PFAS National Leadership Summit," May 22–23, 2017; U.S. Environmental Protection Agency, "EPA's Per- and Polyfluoroalkyl Substances (PFAS) Action Plan," February 14, 2019. For more on Dourson's withdrawal, see Deidre Shesgreen, "Controversial Cincinnati Nominee Asks Trump to Withdraw His Bid for EPA Post," *Cincinnati Enquirer,* December 14, 2017.

CHAPTER 16: WHAT-IFS AND WORST-CASE SCENARIOS

1. According to DuPont records, Sue Bailey had 48,000 parts per trillion PFOA in her blood at the time. Emily had about 300,000 parts per trillion when she got pregnant with Ellie. Because testing methods have evolved considerably over the years, the two numbers might not be directly comparable. Paul Thistleton, DuPont, to S. Takada, Mitsui Fluorochemicals Co. Ltd., "Proposed Employee Blood Sampling Program," memorandum, September 15, 1981, U.S. EPA Public Docket AR226-1390.
2. Garret Ellison, "How Citizen Sleuths Cracked the Wolverine Tannery Pollution Case," *MLive,* June 12, 2019.
3. Shealy Environmental Services, Inc., "Report of Analysis," GZA GeoEnvironmental, Inc., August 15, 2019; Garrett Ellison, "Tannery Chemicals Are 540 Times Above EPA Level in Belmont Well," *MLive,* September 11, 2017; Robert Erb, "Amid PFAS Nightmare, Unlikely Activists Are Born in Michigan," *Bridge Michigan,* June 14, 2019.
4. Rick Snyder, Governor, State of Michigan, "Executive Directive No. 2017-4," November 13, 2017; Michigan PFAS Action Response Team, "Statewide PFAS Survey of Public Water Supplies," accessed in June 2024, www.michigan.gov/pfasresponse/drinking-water/statewide-survey.
5. Keith Matheny, "PFAS Contamination Is Michigan's Biggest Environmental Crisis in 40 Years," *Detroit Free Press,* December 15, 2019.
6. The figure for California is based on the estimated number of state residents who rely on PFAS-tainted water relative to the state's total population. **For more on California,** see U.S. Census Bureau, California, data.census.gov/profile/

California?g=040XX00US06; Arianna Libenson et al., "PFAS-Contaminated Pesticides Applied Near Public Supply Wells Disproportionately Impact Communities of Color in California," *American Chemical Society ES&T Water* 4, no. 6 (May 2024): 2495–2503, doi:10.1021/acsestwater.3c00845. **For New Jersey,** see Rosie Mueller et al., "Quantifying Disparities in Per- and Polyfluoroalkyl Substances (PFAS) Levels in Drinking Water from Overburdened Communities in New Jersey, 2019–2021," *Environmental Health Perspectives* 132, no. 4 (April 2024), doi.org/10.1289/EHP12787. **For nationwide figures,** see Kelly L. Smalling et al., "Per- and Polyfluoroalkyl Substances (PFAS) in United States Tapwater: Comparison of Underserved Private-Well and Public-Supply Exposures and Associated Health Implications," *Environment International* 178, art. 108033 (August 2023), doi.org/10.1016/j.envint.2023.108033.

7. The median blood level for all Hoosick Falls residents sampled between February and April 2016 was 28,300 parts per trillion; for those on the public water supply, it was 55,800. By contrast, the median level among the general U.S. population was about 2,000 parts per trillion at the time. See New York State Department of Health, "Hoosick Falls Area PFOA Biomonitoring Group-Level Results," information sheet, June 2018.

CHAPTER 17: WALL OF RESISTANCE

1. New York State Department of Health, "Drinking Water Quality Council Recommends Nation's Most Protective Maximum Contaminant Levels for Three Unregulated Contaminants in Drinking Water," news release, December 18, 2018.
2. John Kindschuh and Thomas Lee, "State-by-State Regulation of Per- and Polyfluoroalkyl Substances (PFAS) in Drinking Water," *JD Supra,* July 16, 2019; National Conference of State Legislatures, "Per- and Polyfluoroalkyl Substances (PFAS): State Legislation and Federal Action," last updated March 23, 2023, www.ncsl.org/environment-and-natural-resources/per-and-polyfluoroalkyl-substances; Safer States, "Policies for Addressing PFAS," www.saferstates.org/priorities/pfas/.
3. Melissa Nann Burke, "Kildee, House Lawmakers Launching PFAS Task Force," *Detroit News,* January 23, 2019.
4. See, for example, U.S. Congress, Senate, Armed Services Committee, National Defense Authorization Act for Fiscal Year 2020, 116th Congress, S. 1790; U.S. Congress, House, Armed Services Committee, National Defense Authorization Act for Fiscal Year 2020, 116th Congress, H.R. 2500; U.S. Congress, House, Energy and Commerce Committee, Toxic PFAS Control Act, 116th Congress, H.R. 2600.
5. U.S. Congress, House, Energy and Commerce Committee, *Protecting Americans at Risk of PFAS Contamination & Exposure: Testimony Before the Subcommittee on Environment and Climate Change,* 116th Cong. (2019) (testimony of Emily Marpe).

6. Debbie Dingell et al., to Adam Smith, Chairman, House Armed Services Committee, James M. Inhofe, Chairman, Senate Committee on Armed Services, Mac Thornberry, Ranking Member, House Armed Services Committee, and Jack Reed, Ranking Member, Senate Committee on Armed Services, October 22, 2019.
7. Food & Water Watch, "PFAS and the Chemistry of Concealment," November 7, 2023; Open Secrets, "Client Profile: 3M Co," 2019–2020; Open Secrets, "3M Co Profile," 2020 cycle; Open Secrets, "PAC Profile: 3M Co," 2018–2020; Christopher Magan, "Campaign Spending Watchdog Says 3M Donated Illegally," *Pioneer Press* (St. Paul, MN), March 8, 2019.
8. Annie Snider, "Inside a Corporate Giant's Fight to Thwart a Massive Pollution Tab," *Politico,* November 23, 2018; John Butenhoff, Medical Department, 3M, to Michael A. Santoro et al., "Notes on Proposal to NTP," September 27, 2004, State of Minnesota v. 3M Co., Plaintiff's Exhibit 1966; John L. Butenhoff, "Toxicological Research Program in Perfluorinated Chemistries," slide presentation, March 29, 2008, State of Minnesota v. 3M Co., Plaintiff's Exhibit 2206.
9. Notably, the coalition's briefing materials relied partly on an ethically questionable 2018 study involving forty-nine cancer patients who were exposed to large doses of PFOA, supposedly to test its therapeutic potential. The treatment failed to slow the growth of their tumors, but because the cancer patients metabolized the chemical differently than the subjects in other studies, industry groups seized on the data to sow doubt about the science on PFOA's health effects and longevity in the human body. **On the Responsible Science Policy Coalition:** Open Secrets, "Client Profile: Responsible Science Policy Coalition," 2019; Responsible Science Policy Coalition, "Polyfluoroalkyl Substances: Best Practices for Science Policy Decisions," slide presentation to the Conference of Western Attorneys General, July 24, 2018; Responsible Science Policy Coalition, "Our nation's leaders are asking the right questions on PFAS. Industry stands ready to answer those questions . . . ," undated congressional briefing document. **On the cancer-patient study:** Sharon Lerner, "Industry Cites 3M Research on Cancer Patients Exposed to PFOA to Claim the Chemical Isn't So Bad," *The Intercept,* August 12, 2019; Matteo Convertino et al., "Stochastic Pharmacokinetic-Pharmacodynamic Modeling for Assessing the Systemic Health Risk of Perfluorooctanoate (PFOA)," *Toxicological Sciences: An Official Journal of the Society of Toxicology* 163, no. 1 (May 2018): 293–306, doi:10.1093/toxsci/kfy035.
10. Eric Lipton and Julie Turkewitz, "Pentagon Pushes for Weaker Standards on Chemicals Contaminating Drinking Water," *New York Times,* March 14, 2019; Food & Water Watch, "PFAS and the Chemistry of Concealment"; Open Secrets, "Bills Lobbied by American Petroleum Institute, 2019," last updated on July 24, 2024.
11. Hannah Northey and Ariana Figueroa, "Intense PFAS Lobbying Pits Greens Against Water Utilities," *E&E News by Politico,* June 11, 2020; American Water Works Association et al. to individual members of the U.S. House of

Representatives, "Re: Opposition to H.R. 535, the PFAS Action Act," January 8, 2020.

12. U.S. Congress, House Oversight and Reform Committee, *The Devil They Knew: PFAS Contamination and the Need for Corporate Accountability, Part II,* 116th Cong. (2019) (testimony of Daryl Roberts, DuPont de Nemours, Inc.; Paul Kirsch, Chemours Company; and Denise Rutherford, 3M Company); The Chemours Company v. DowDuPont Inc., C.A. No. 2019-0351-SG (2019).
13. Gabe Schneider, "3M Grilled over PFAS Chemicals at Congressional Hearing," *MinnPost,* September 11, 2019.
14. *The Devil They Knew: PFAS Contamination and the Need for Corporate Accountability, Part II* (testimony of Robert Bilott, Taft Stettinius & Hollister LLP).
15. Ariana Figueroa, "How PFAS Negotiations Fell Apart," *E&E News by Politico,* December 9, 2019; Todd Spangler, "Trump Administration, GOP Strip Out PFAS Standard, Cleanup Requirements from Defense Bill," *Detroit Free Press,* December 10, 2019.
16. World Health Organization, "Statement on the Second Meeting of the International Health Regulations (2005) Emergency Committee Regarding the Outbreak of Novel Coronavirus (2019-nCoV)," January 30, 2020.
17. Jonathan Allen, Phil McCausland, and Cyrus Farivar, "Behind Closed Doors, Trump's Coronavirus Task Force Boosts Industry and Sows Confusion," *NBC News,* April 11, 2020.
18. DuPont, "DuPont™ Tyvek® Activates Operation Airbridge to Speed Up PPE Supply for COVID-19 Response," news release, April 7, 2020.
19. Amy Brittain and Isaac Stanley-Becker, "Senators Seek Investigation into Project Airbridge Deliveries of Protective Medical Gear," *Washington Post,* June 9, 2020; Office of Inspector General of the Department of Homeland Security, "FEMA Did Not Provide Sufficient Oversight of Project Airbridge," February 7, 2023; DuPont, "DuPont™ Tyvek® Activates Operation Airbridge"; Jonathan Allen, Phil McCausland, and Cyrus Farivar, "Silent Partner in Coronavirus Contract Sold Protective Gear to U.S. for Double the Cost," *NBC News,* April 19, 2020.

CHAPTER 18: VICTORY

1. Decision denying defendants' motion to dismiss, Baker v. Saint-Gobain Performance Plastics Corp., May 18, 2020.
2. For the Petersburg case, see Weitz & Luxenberg, "Decision Paves the Way for Trial in Petersburgh, New York Water Contamination Case," news release, February 6, 2020. For the Bennington case, see Jim Therrien, "OSHA Filing Against Saint-Gobain Could Impact Lawsuit," *Bennington (VT) Banner,* April 15, 2021.
3. Complaint, Amiel Gross v. Saint-Gobain Corp. et al., U.S. Department of

Labor, Occupational Safety & Health Administration, no. 2021-SDW-00001, April 6, 2021.

4. Complaint, Gross v. Saint-Gobain Corp.
5. According to Saint-Gobain, Gross was fired for "blatantly and repeatedly" acting "in an insubordinate manner" and violating unspecified company policies. See Decision and Order of Remand, Gross v. Saint-Gobain Corp., Department of Labor, Administrative Review Board, no. 2022-005, April 18, 2022, 11.
6. Complaint, Gross v. Saint-Gobain Corp; Therrien, "OSHA Filing"; Jim Therrien, "Settlement Agreement Reached in Bennington PFOA Suit," *Bennington (VT) Banner,* November 11, 2021.
7. Class Settlement Agreement, Baker v. Saint-Gobain Performance Plastics Corp., July 21, 2021.
8. Brendan J. Lyons, "$65 Million Settlement Filed in Hoosick Falls PFOA Water Contamination," *Times-Union* (Albany, NY), July 21, 2021.
9. For more information about the settlement that Taconic Plastic reached with Petersburgh residents, see Class Settlement Agreement, Burdick v. Tonoga Inc., October 1, 2021.
10. New York Department of Environmental Conservation, "DEC Announces $45 Million Agreement with Saint-Gobain, Honeywell for Hoosick Falls Water Supply and Payment of State Emergency Response Costs," news release, May 12, 2023; NYDEC, "DEC Releases Final Plan to Provide New Permanent Clean Drinking Water Source for Hoosick Falls," news release, December 3, 2021.
11. A comprehensive 2019 Federal Trade Commission study found that the median claims rate for consumer class-action suits was just 9 percent. While there are no comparable figures for environmental cases, participation rates for class-action settlements generally are known to be nominal. See Federal Trade Commission, "Consumers and Class Actions: A Retrospective and Analysis of Settlement Campaigns," September 2019; Mayer Brown LLP, "Do Class Actions Benefit Class Members? An Empirical Analysis of Class Actions," December 11, 2013.
12. Class Settlement Agreement, Baker v. Saint-Gobain Performance Plastics Corp.

CHAPTER 19: TO THE ENDS OF THE WORLD

1. In addition to the media coverage, some prominent politicians went out of their way to honor the beloved doctor and his fight for clean drinking water. In May 2023, for instance, Senate Majority Leader Chuck Schumer sent the Martinezes a letter explaining how a 2016 conversation with Marcus had inspired him to fight for provisions, like federal caps on PFAS in drinking water, that were then on the verge of being enacted. "The word 'hero' gets tossed around quite a bit in our vernacular," Schumer wrote. "But it seems to me that the real

heroes I have encountered in my life have been the ordinary among us who, in times of crisis, have done the extraordinary. By that definition, Dr. Marcus Martinez is a true hero." Charles E. Schumer, United States Senator, to Marcus Martinez et al., May 13, 2023.

EPILOGUE

1. Ian T. Cousins et al., "Outside the Safe Operating Space of a New Planetary Boundary for Per- and Polyfluoroalkyl Substances (PFAS)," *Environmental Science & Technology* 56, no. 16 (August 2022): 11172–11179, doi:10.1021/acs.est.2c02765.
2. See, for example, Kit Granby et al., "Per- and Poly-Fluoroalkyl Substances in Commercial Organic Eggs Via Fishmeal in Feed," *Chemosphere* 346, art. 140553 (January 2024), doi:10.1016/j.chemosphere.2023.140553; Heather Schwartz-Narbonne et al., "Per- and Polyfluoroalkyl Substances in Canadian Fast Food Packaging," *Environmental Science & Technology Letters* 10, no. 4 (March 2023): 343–349, doi:10.1021/acs.estlett.2c00926; Kathryn M. Rodgers et al., "How Well Do Product Labels Indicate the Presence of PFAS in Consumer Items Used by Children and Adolescents?," *Environmental Science & Technology* 56, no. 10 (May 2022): 6294–6304, doi:10.1021/acs.est.1c05175.
3. Tom Perkins, " 'I Don't Know How We'll Survive': The Farmers Facing Ruin in Maine's 'Forever Chemicals' Crisis," *Guardian*, March 22, 2022; Hiroko Tabuchi, "Something's Poisoning America's Land: Farmers Fear 'Forever' Chemicals," *New York Times*, August 31, 2024; Nadia Barbo et al., "Locally Caught Freshwater Fish Across the United States Are Likely a Significant Source of Exposure to PFOS and Other Perfluorinated Compounds," *Environmental Research* 220, no. 9 (December 2022), doi:10.1016/j.envres.2022.115165.
4. Ashley M. Lin et al., "Landfill Gas: A Major Pathway for Neutral Per- and Polyfluoroalkyl Substance (PFAS) Release," *Environmental Science & Technology Letters* 11, no. 7 (July 2024): 730–737, doi:10.1021/acs.estlett.4c00364; Cheryl Hogue, "Incinerators May Spread, Not Break Down PFAS," *Chemical & Engineering News*, April 28, 2020; S. Björklund, "Exploring the Occurrence, Distribution and Transport of Per- and Polyfluoroalkyl Substances in Waste-to-Energy Plant," PhD diss., Umea University, Sweden, 2024.
5. Bo Sha et al., "Constraining Global Transport of Perfluoroalkyl Acids on Sea Spray Aerosol Using Field Measurements," *Science Advances* 10, no. 14 (April 2024), doi:10.1126/sciadv.adl1026.
6. Isabelle J. Neuwald et al., "Ultra-Short-Chain PFASs in the Sources of German Drinking Water: Prevalent, Overlooked, Difficult to Remove, and Unregulated," *Environmental Science & Technology* 56, no. 10 (May 2022): 6380–6390, doi:10.1021/acs.est.1c07949.
7. **On beer:** Marco Scheurer et al., "Ultrashort-Chain Perfluoroalkyl Substance Trifluoroacetate (TFA) in Beer and Tea: An Unintended Aqueous Extraction," *Food Chemistry* 351, art. 129304 (July 2021), doi:10.1016/j.foodchem.2021.129304. **On bottled water:** see Pesticide Action Network Europe, "TFA:

The Forever Chemical in the Water We Drink," September 2024. **On baby food:** Patrick van Hees et al., "Trifluoroacetic Acid (TFA) and Trifluoromethane Sulphonic Acid (TFMS) in Juice and Fruit/Vegetable Purees," Eurofins Food & Feed Testing Center, September 2024. **On groundwater and rivers:** Pesticide Action Network Europe, "TFA in Water: Dirty PFAS Legacy Under the Radar," May 2024.

8. Guomao Zheng et al., "Elevated Levels of Ultrashort- and Short-Chain Perfluoroalkyl Acids in US Homes and People," *Environmental Science & Technology* 57 (October 2023): 15782–15793, doi:10.1021/acs.est.2c06715.
9. European Chemicals Agency, "Annex to the Annex XV Restriction Report, Proposal for a Restriction," March 20, 2023, 156–170.
10. There is one treatment technology that effectively removes ultra-short-chain PFAS: reverse osmosis. However, it is not feasible to deploy on a large scale because of the exorbitant cost.
11. Neuwald et al., "German Drinking Water."
12. Jeffrey Kluger, " 'Forever Chemical' Lawsuits Could Ultimately Eclipse the Big Tobacco Settlement," *Time,* July 12, 2023; Hiroko Tabuchi, "Lawyers to Plastics Makers: Prepare for 'Astronomical' PFAS Lawsuits," *New York Times,* April 8, 2024; Adam Piore, " 'PFAS . . . everywhere': A Mass. class-action lawsuit may set a new standard for damages," *Boston Globe,* September 14, 2024.
13. U.S. Environmental Protection Agency, "Biden-Harris Administration Finalizes First-Ever National Drinking Water Standard to Protect 100M People from PFAS Pollution," news release, April 10, 2022; U.S. Environmental Protection Agency, PFAS National Primary Drinking Water Regulation Rulemaking, 40 CFR Parts 141 and 142, March 13, 2023.
14. Even if the federal drinking-water standards do survive, the EPA may not be able to enforce them since the 2024 Supreme Court ruling overturning the Chevron Doctrine has significantly curtailed the agency's power.
15. Amara's Law, Minnesota Statute 116.943 (2023); Amudalat Ajasa, "She Died Fighting 'Forever Chemicals.' They Still Linger in Her Town," *Washington Post,* July 14, 2023; "Policies for Addressing PFAS," accessed in October 2024, www.saferstates.org/priorities/pfas/.
16. Apple Inc., "Apple's Commitment to Phasing Out Per- and Polyfluoroalkyl Substances (PFAS)," November 2022; Toxic-Free Future, "Retailers Committing to Phase Out PFAS as a Class in Food Packaging and Products," accessed in April 2024, www.toxicfreefuture.org/mind-the-store/.
17. 3M Company, "3M Announces Exit from PFAS Manufacturing," news release, December 20, 2022.
18. Martin Scheringer et al., "Is a Seismic Shift in the Landscape of PFAS Uses Occurring?," *Environmental Science & Technology* 58, no. 16 (April 2024): 6843–6845, doi:10.1021/acs.est.4c01947; Investor Initiative on Hazardous Chemicals, "Excerpts from Investor Letter Sent by the Members of the Investor Initiative on Hazardous Chemicals," November 15, 2023.
19. Scheringer et al., "Is a Seismic Shift in the Landscape of PFAS Uses Occurring?"; Chemours Company, "Protecting Fluoropolymers and Fluorinated

Gases, the Building Blocks of our Future," www.chemours.com/en/pfas-advocacy.

20. Sharon Lerner, "Chemours Lobbied EPA to Avert Use of Natural Refrigerants," *The Intercept,* August 27, 2018; U.S. Environmental Protection Agency, "National PFAS Testing Strategy: Identification of Candidate Per- and Polyfluoroalkyl Substances (PFAS) for Testing," October 2021, 5.
21. Merritt Wallick, "The Reluctant Environmentalists," *News Journal* (Wilmington, DE), August 26, 1991; Brigitte Smith, "Ethics of Du Pont's CFC Strategy 1975–1995," *Journal of Business Ethics* 17, no. 5 (April 1998): 557–568.
22. Seth Cagin, *Between Earth and Sky: How CFCs Changed Our World and Endangered the Ozone Layer* (New York: Pantheon, 1993), 309.
23. The 2000 deadline was for developed countries; developing countries were given an additional ten years. Smith, "Du Pont's CFC Strategy"; United Nations, "Amendment to the Montreal Protocol on Substances that Deplete the Ozone Layer," *Treaty Series*, vol. 1598, June 29, 1990.
24. Steve Seidel et al., "Technological Change in the Production Sector Under the Montreal Protocol," Center for Climate and Energy Solutions, October 2015; United Nations, "Climate Change: Ozone Layer Still Well on Track for Full Recovery," *UN News,* September 16, 2024.
25. Ian T. Cousins et al., "The Concept of Essential Use for Determining When Uses of PFASs Can Be Phased Out," *Environmental Science: Processes & Impacts* 21, no. 11 (June 2019): 1803–1815, doi:10.1039/c9em00163h; Ian T. Cousins et al., "Finding Essentiality Feasible: Common Questions and Misinterpretations Concerning the 'Essential-Use' Concept," *Environmental Science: Processes & Impacts* 23, no. 8 (June 2021): 1079–1087, doi:10.1039/d1em00180a.
26. European Commission, "Commission Defines Principles on Limiting Most Harmful Chemicals to Essential Uses," news release, April 22, 2014.
27. John Gardella, "L'Oreal PFAS Lawsuit Shows the Danger of ESG Marketing," *National Law Review,* March 14, 2022; Patrick Ambrosio, "Snapple 'All Natural' Iced Tea Label Suit Survives Dismissal Bid," *Bloomberg Law,* October 14, 2024; Jonathan Stempel, "Trojan Condoms Contain 'Forever Chemicals,' Lawsuit Claims," *Reuters,* September 10, 2024.

Acknowledgments

This book would not exist if it weren't for the courage and generosity of the people of Hoosick Falls. I am enormously grateful to Michael Hickey, Emily Marpe, and Marcus Martinez for opening their lives to me and answering my many questions over eight difficult years.

I am also deeply indebted to Heather Allen, Rob Allen, David Bond, Brian Bushner, Angela Cottrell Carknard, Judith Enck, David Engel, Loreen Hackett, Megan Henry, Andrew Hoag, Jeff Hickey, Oliver Hickey, Rhonda Hickey, Sue Hickey, Philip Leonard, Katy Lilac, Gloria Martinez, Gretchen Johnson-Martinez, Jamie Martinez, Philip Martinez (aka "Old Doc"), Gwen Young, and Michelle Vandermark.

Numerous other people informed my understanding of the events in Hoosick Falls and neighboring towns. Among them: Kevin Allard, Mike Bacon, Ben Barton, Alexandra Becker, Beatrice Berle, David Borge, Cynthia Brewster, Jane Conte, Jim Cusack, Jamie Davis, Cathy Dawson, Tami Duket, David Galusha, Jim Goodine, Annette Griffith, Ameil Gross, David Hassel, Richard Hickey, Shaina Kasper, Helen Kennedy, Brendan J. Lyons, Derek Miller, James Monahan, Carol Moore, Kevin O'Malley, Robert Ridley, Suzanne

Seymour, Richard Spiese, Alyssa Schuren, Sandy Sumner, Patrice Zedalis, Thomas Zelker, and Marianne Zwicklbauer.

Many thanks to them and to the whistleblowers, activists, and community residents who educated me on the PFAS crises in other parts of the country. I am especially grateful to Andrea Amico, Bucky Bailey, Sue Bailey, Earl Botkin, Paul Brooks, Tracy Carluccio, Robert Delaney, Emily Donovan, Bob Griffin, Darlene Kiger, Joe Kiger, Sheila Lowther, Callie Lyons, Eileen Murphy, Fardin Oliaei, Della Tennant, Jim Tennant, Kenton Wamsley, and Sandra Wynn-Stelt.

Every book is, to some degree, a group endeavor. This one wouldn't have been remotely possible without the many patient scientists and scholars who helped me understand the complex world of PFAS and other hormone disrupting chemicals. Thanks to Hans Peter Arp of the Norwegian Geotechnical Institute and the Norwegian University of Science and Technology; Linda Birnbaum of the National Institute of Environmental Health Sciences (retired); Phil Brown of Northeastern University; Courtney Carignan of Michigan State University; Richard Clapp of the University of Massachusetts Lowell; Alissa Cordner of Whitman College; Ian T. Cousins of Stockholm University; Jamie DeWitt of Oregon State University; Alan Ducatman of West Virginia University (emeritus); Philippe Grandjean of Harvard University and the University of Southern Denmark; Evan Hepler-Smith of Duke University; Christopher Higgins of the Colorado School of Mines; Xindi C. Hu of George Washington University; Andrew Lindstrom of the EPA's National Exposure Laboratory (retired); Detlef Knappe of North Carolina State University; Lauren Richter of the University of Toronto; Graham Peaslee of the University of Notre Dame; Laurel A. Schaider of the Silent Spring Institute; Laura N. Vandenberg of the University of Massachusetts Amherst; and Frederick S. vom Saal of the University of Missouri.

During my research, I also relied heavily on archival collections, including the Bush-Conant Files Relating to the Development of the

Atomic Bomb and the Papers of Lt. Gen. Leslie R. Groves, Jr., at the U.S. National Archives and Records Administration; the Robert A. Kehoe Archival Collection at the University of Cincinnati's Henry R. Winkler Center for the History of the Health Professions; the William T. Miller Papers at Cornell University Library's Division of Rare and Manuscript Collections; the Christopher Bryson and Joel Griffiths Papers at the University of Massachusetts Amherst; the James J. Bohning Collection at the Science History Institute in Philadelphia; the W. C. Hueper Papers at the National Library of Medicine in Bethesda, Maryland; the Stephane Groueff Collection at Boston University's Howard Gotlieb Archival Research Center; and both the Crawford H. Greenewalt Papers and E.I. du Pont de Nemours & Company Atomic Energy Division Records at the Hagley Museum and Library in Wilmington, Delaware.

I owe a special debt of gratitude to Melanie Benesh and Scott Faber of the Environmental Working Group for answering my policy questions, and to the devoted lawyers who guided me through the thicket of PFAS litigation. Rob Bilott provided a wealth of valuable information, as did James Bilsborrow of Weitz & Luxenberg; Harry Deitzler of Hill, Peterson, Carper, Bee & Deitzler; Ned McWilliams of Levin Papantonio; and Stephen G. Schwarz of Faraci Lange.

Knitting these varied strands together was a hugely daunting task. I never could have done it without my brilliant editor, Amanda Cook, whose keen editorial vision has shaped every facet of the book, from the outline to the final draft. She has given far more to it than I had the right to expect. I'd also like to thank the rest of the Crown team, including publicist Penny Simon, marketer Chantelle Walker, and assistant editor Katie Berry, who diligently shepherded the project through its many phases.

Notably, this book grew out of a 2015 article in *HuffPost Highline,* which was edited by my exceptionally talented friend, Rachel Morris, and was later nominated for a National Magazine Award.

Rachel has also served as a consulting editor on the book itself. It is thanks largely to her prose wrangling and thoughtful guidance on issues like structure that I was able to transform this complicated, technical story into a nuanced narrative.

My work has also benefited from the deft advice of several early readers. Rebecca Altman, Benny Peterson, Kate Rodemann, Ruth Shorter, and Allison Sumner, thank you for making this book better. Thanks also to my talented research assistants—Carson Kessler, Jorga Rose, Justin Shea, Nick Tabor, and especially Matthew Giles—for working to ensure the accuracy of this complex text.

To the authors whose work shaped my thinking about the themes in this book—Erik M. Conway, Devra Davis, Dan Fagin, Eliza Griswold, Patrick Radden Keefe, David Michaels, Gerald Markowitz, Naomi Oreskes, David Rosner, and Leonardo Trasande—I am honored to be standing on your shoulders.

A heartfelt thanks to my agent, Larry Weissman, for helping me to find a home for this book, and to Ann Marie Lipinski, the curator of the Nieman Foundation for Journalism at Harvard University. It was under her tutelage that I incubated many of the ideas for this project and gathered the confidence to see it through.

Finally, Andreas and Cody, thank you for standing by me and helping me find balance throughout this consuming journey and all the other journeys we've shared. You are my why.

Index

About the Author

Mariah Blake is an investigative journalist whose writing has appeared in *The Atlantic*, *Mother Jones*, *The New Republic*, and other publications. She was a Murrey Marder Nieman Fellow in Watchdog Journalism at Harvard University.

@MARIAHCBLAKE (X)

P.O. 0005607322 20250311